D0992459
R0006616817

A System Approach to Water Distribution Modeling and Control

A System Approach to Water Distribution Modeling and Control

Robert DeMoyer, Jr.
General Electric Company

Lawrence B. Horwitz
Polytechnic Institute of
New York

Lexington Books
D.C. Heath and Company
Lexington, Massachusetts
Toronto London

Library of Congress Cataloging in Publication Data

DeMoyer, Robert.
A system approach to water distribution modeling and control.

Bibliography: p.
Includes index.
1. Waterworks—Mathematical models. 2. Waterworks—Automation. I. Horwitz, Lawrence B., joint author. II. Title.
TD487.D45 628.1'44 74-33803
ISBN 0-669-99127-9

Published simultaneously in Canada

Printed in the United States of America

International Standard Book Number: 0-669-99127-9

Library of Congress Catalog Card Number: 74-33803

To Lois and to Karen, Toby, and Sara

Contents

List of Figures

List of Tables

Preface

For many years, urban and suburban water departments have been successfully and reliably pumping potable water from a variety of sources to residential and commercial customers. However, two recent events have had an impact on the continuing development of water distribution technology. Most water departments have been approached by computer manufacturers describing the benefits to be gained by the application of data logging or process control computers to aid in water distribution. While many computers are now in use, there are no generally accepted algorithms for water distribution pump control. More recently, there has been a national concern for the reduction of energy consumption. Energy surcharges have significantly increased the already considerable electrical cost involved in most distribution pumping. The current motivation for water distribution control research is clear: How can the volumes of operating data collected by an on-line computer be used to reduce distribution pumping energy consumption with no decrease in reliability?

While one of the goals of this book is to describe in detail a particular cost reducing computer control algorithm, there are perhaps two more fundamental goals. One is to extol the virtues of the systems approach and the use of some of the many systems analysis tools. Some of the tools used here range from statistical modeling and Fourier analysis—familiar to hydraulogists—to Kalman filtering and dynamic programming, which have in the past rarely been associated with hydraulics or water distribution. The second fundamental goal is to emphasize the importance of recognizing that the distribution system is driven by a stochastic forcing function —that is, customer demand. Many engineers involved in water distribution are not accustomed to adding probabalistic considerations to their analysis. Some pump control algorithms developed without such analysis have been subject to excess pump cycling as a result of the random component in customer demand. Probabilistic considerations are prominent in each of the major topics discussed in this text.

How can pumping energy consumption be reduced? Since the answer to this question varies according to the nature of the district being served, for the present study, the districts under consideration have been confined to those that are largely residential and that contain elevated storage tanks. Energy consumption in such a district can be changed by modifying the timing involved in forcing water into elevated storage. Tank depth policy is, of course, subject to tank depth constraints and customer pressure constraints. While most analytical control theory is considerably complicated by the presence of constraints, the computations required by dynamic programming control are actually reduced. For this reason, and others described in Chapter 5, a control algorithm was developed based on dynamic programming.

For dynamic programming control to be practical, a system model is needed that requires orders of magnitude less computation time than that required by a traditional full network model, which is based on the iterative solution of a large set of nonlinear differential equations. For this reason, a statistical model was developed, as derived in Chapter 2. Not only is the model fast, but it overcomes the costly and difficult problem of availability of accurate pipe parameters required by a full network model. The statistical model is developed from operating data provided by a data logging computer, and, for this reason, can be readily adapted to changing conditions. Chapter 3 describes the use of a modified Kalman filter to perform the recursive parameter estimation needed to keep the statistical model up to date without programming modifications.

Finally, dynamic programming control requires the prediction of time varying demand. In Chapter 4, Fourier analysis and related statistical tests permit the separation of the predictable component of time varying demand from the unpredictable random component.

Control of a district in Philadelphia, Pennsylvania, has been simulated with a resultant savings of 4.9 percent in energy consumption, relative to current operating policy. Significantly, there is an even greater saving in comparison to the frequently advocated "floating tank" policy.

Acknowledgments

This research could not have been carried out without the help and cooperation of many people at the Philadelphia Water Department. Commissioner Carmen F. Guarino has been a constant advocate of distribution research. Gratitude is also expressed to Joseph V. Radziul, Chief, Research and Development, to Victor Pagnotto, Chief, Water Operations; and to Andrew Peters, Chief, Load Control.

The General Electric Company is to be acknowledged for its support, as is the U. S. Department of Interior, Office of Water Resources Research, for its sponsorship of related water distribution research.

A System Approach to Water Distribution Modeling and Control

1 Introduction

With the continuing trend of population shifting to large urban and suburban areas, the demands upon water distribution systems are growing. The increased size and complexity of water distribution systems are making economical manual control of pumping stations difficult, so attention is turning to automatic control. The water industry lags behind the industrial giants in automation progress because of past policies of "too cheap water" [68], which implies that, in the past, when determining a price for water, the industry did not allow for enough research and development.

Several large city water distribution systems are in various stages of development of automatic control. The emphasis has been to provide adequate service through computer implemented logic sequences built up from the composite experience of many system operators. No true optimization study has been undertaken, and even the proposed philosophy of how economical control should be implemented varies considerably. Considering the fact that large city power costs for pumping can exceed $1 million annually, an increase in pumping efficiency is desirable [58, 64].

The objective here is to present an overall system study of the automatic control of elevated-storage-dominated water distribution systems. Existing distribution systems are to be controlled, with no modification to hardware, in a way that will ensure that adequate customer pressure is maintained. Operating costs are to be minimized, while at the same time various system variables are kept within constraints. To carry out these objectives, concepts of modeling and control—new to water distribution—have been developed.

The Water Distribution System[a]

A water distribution system has been defined as "all water works components for the distribution of finished or potable water by means of gravity storage or pumps through distribution piping networks to customers or other users, including distribution equalizing storage" [58].

Sources of potable water include wells, filtration plants, and large transmission pipelines. Filtration plants often feed small buffering reser-

[a]See bibliographical notes 12, 21, 70, and 86.

voirs so that demand variations do not upset the filtration process. From such sources, water is forced into the pipeline distribution network by centrifugal pumps. Water is distributed to customers through pipes varying in size of several feet in diameter to customer service pipes on the order of one inch. Larger distribution pipes form a network of closed loops, with "trees" being formed only in the smaller local pipes. The fact that the network contains many closed loops makes it possible to valve off breaks without interrupting service to many customers.

The normal pressure range within a network is from 30 to 70 pounds per square inch (psi), but, in hilly districts, the pressure can range up to 120 psi. Large distribution systems are usually divided into several pressure districts that remain largely independent, though they can be interconnected to meet emergencies. Pressure districts having relatively high elevations are called high service districts, while those at lower elevations are low service districts.

Most pressure districts contain one or more elevated storage tanks whose primary function is to equalize pumping. When pumping capacity on line exceeds demand, these tanks fill; then, during periods of high demand, the tanks empty. Tanks perform the function of stabilizing system pressure in districts where only a few discrete pump sizes are available. A distribution system without elevated storage would require a larger number of pumps to keep pressures within given bounds than would a similar system containing elevated storage. Elevated storage also provides a reserve to meet fire or emergency demands. The pressures in a district are characterized by a hydraulic grade line, which is, by definition, made up of the elevations to which water would rise in open vertical tubes connected to various points within the network. These elevations, expressed in feet, are referred to as *heads*. Elevated tanks stabilize the hydraulic grade line so that it changes little when pump changes are made. On the other hand, when there is no equalizing storage, a pump change has a noticeable effect on the hydraulic grade line, and, hence, on customer pressures. The former system might be called tank dominated, while the latter is pump dominated.

The effect of pump changes with and without elevated storage can be shown by head-flow curves and hydraulic grade lines (HGLs) for a simple system containing one pump station. Figures 1-1 and 1-2 illustrate a system with and without equalizing storage. The flow through a pump is determined by a supply and demand interaction. Pumps have negatively sloped supply curves, which show that increased head across the pump decreases flow through it. The system demand curve has a positive slope, indicating that an increased head is necessary to force more water into the system. The operating point is determined by the intersection of supply and demand. In a system with equalizing storage, the demand system curve rises

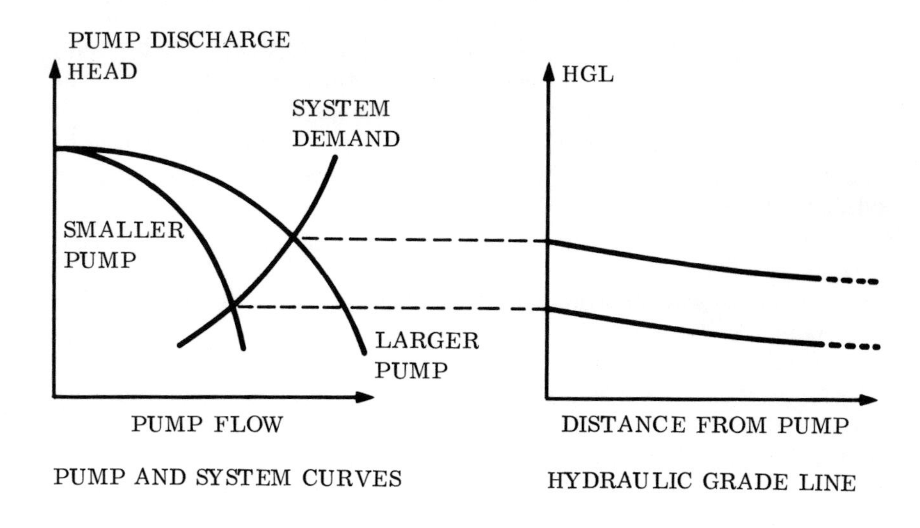

Figure 1-1. Equalizing Storage

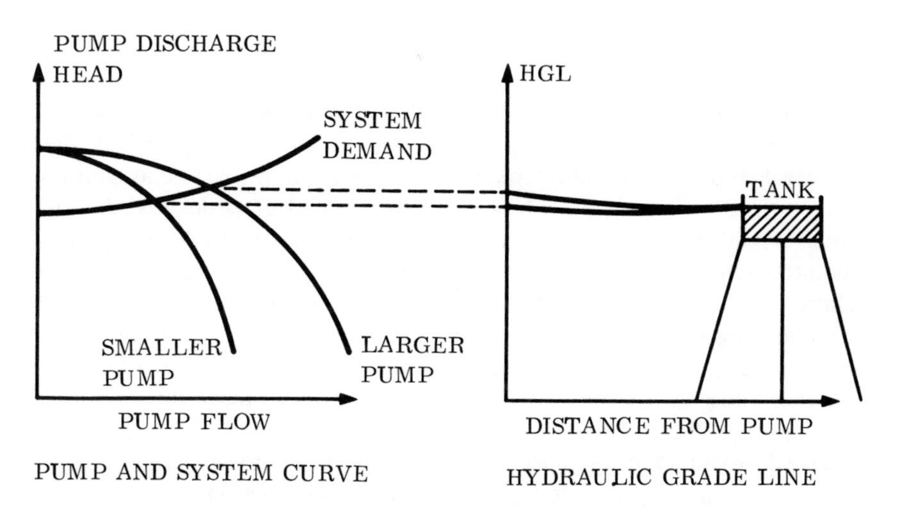

Figure 1-2. Direct Pumping

and falls as tank depths change, while the shape changes little. For direct pumping, the curve shifts horizontally with varying demand.

Once the pressure range permissible at customer services has been established, there is little control to be done in direct pumping. On the other hand, when there is equalizing storage, the economy of operation is partly a function of the timing of when water is forced into storage and when it is drawn off. From this point on, only elevated-storage-dominated pressure districts will be discussed. In such districts, control is effected by putting discrete pumps on and off at one or more pumping stations. Furthermore, it will be assumed that limits in tank depths that lead to adequate pressures and that provide emergency reserves have been established and are considered as given.

Current Methods of Modeling and Analyzing Water Distribution Systems

Current methods of modeling and analysis are primarily oriented toward design rather than control. For the purpose of design, analog or digital analysis of a distribution system is essential to determine the pressure and flow effects resulting from proposed piping changes and additions. A brief review of existing modeling and analysis techniques will show their advantages and disadvantages with respect to control.

Full Network Solutions

A full network solution consists of the calculation of all pipe flows and node pressures. In order to make the analysis practical, a network is first "skeletonized" such that all pipes under six to eight inches in diameter are deleted, and all loads (takeoffs) are concentrated at nodes. A network model is said to be balanced when the net flow into any node is zero, and when the net head drop (change in energy) around any loop is zero. These balance requirements are the hydraulic equivalent of Kirchhoff's laws. The problem formulation is the same as in the electrical problem, but because the head drop in a section of pipe is a nonlinear function of flow, iterative digital techniques are required, where the number of iterations increases with increased accuracy.

Head loss in a pipe is approximately proportional to the square of pipe flow, so for the purpose of a simplified analysis, the Darcy equation is frequently used [89]:

$$h_L \propto Fr\, L\, Q^n / D^m \,, \tag{1.1}$$

where h_L represents head loss; Fr, friction factor due to pipe roughness; L, pipe length; D, pipe diameter; Q, flow; $n = 2$; and $m = 5$. In practice, the exponent of flow varies from 1.75 to 2.0. Extensive analysis of data has shown that the empirical Hazen-Williams equation is most widely applicable:

$$h_L \propto LQ^n/C^nD^m, \qquad (1.2)$$

where C is the Hazen-Williams C roughness coefficient; $n = 1.85$; and $m = 4.68$.

Basic Solution Procedures—A Review of Previous Work: In 1936, Hardy Cross [14] presented a digital technique to balance a hydraulic network model. Analogous to the solution of an electrical network, the Hardy Cross method can be formulated either in terms of loops or nodes. When a loop formulation is used, an initial flow solution is required that satisfies flow continuity at all nodes. Then, through use of the method of balanced heads, loop flow corrections are made, one loop at a time, until all loop net head increases approach zero. In a node formulation, the method of balanced flows is used in which node heads are corrected one at a time until all net node flow errors approach zero.

A loop formulation has the advantage of being easier to visualize and has fewer equations involved because there are on the order of 25 percent fewer loops than nodes in a typical distribution network. On the other hand, there are many possible ways to define the loops, and some loop definitions lead to faster convergence than others. Also, as was indicated, an initial feasible flow solution is required. The rapidity of the solution convergence is enhanced if the initial flow solution guess is close to the final flow solution. In either case, convergence is not guaranteed. Recent improvements to the basic Hardy Cross method are primarily directed toward easing the bookkeeping burden on the user and improving speed of convergence. In 1966, Shirley and Bailey [74] presented a loop formulation which includes an automatic initial flow calculation and extra loops to speed convergence. McCormick and Bellamy [49, 1968] presented a node formulation. Their article is computer program oriented, and while the node formulation is difficult to visualize, there is no need for loop definition or initial flow solutions. In a recent publication by Gilman, Goodman, and Metkowski [24, 1972], the loop formulation is improved upon by a good initial flow solution produced by an optimal spanning tree. In this case, the initial flow solution is created by a tree that contacts each node, and forms no closed loops. The tree is formed such that more water flows through larger pipes than through smaller pipes, which results in an initial flow guess that is close to the final balanced solution.

Loop or node equations can all be simultaneously corrected by Newton's method. This method, originally formulated by Martin and Peters [46, 1963], requires the inversion of a matrix, equal in size to the number of loops or nodes, at each iteration. In 1968, Shamir and Howard [73] extended the Newton-Raphson method to solve for unknown heads, consumptions, and pipe resistances. This development is not possible with the Hardy Cross method. However, the Newton-Raphson method generally requires more computer storage than Hardy Cross. In 1970, Epp and Fowler [20] developed a loop formulation that includes automatic loop numbering leading to a banded matrix, the majority of whose terms fall close to the main diagonal, which is easy to invert numerically. The use of pseudoloops permits inclusion of pumps and reservoirs. Similarly, Lam and Wolla [20, 40, 41, 1972] produced computer generated node equations by use of graph theory. Michel and Wolfner [63, 1972] present a list of commercially available network solution programs. In addition, they describe a large capacity loop formulation program of their own based on the work of Epp and Fowler.[b]

Networks can also be analyzed in an analog fashion. The Mellroy [50, 1950] analyzer employs "fluistors" whose voltage drop is proportional to current to the 1.85 power. These analyzers are large and cumbersone to set up. While most work is digital, Wood [89, 1971] has described network analysis by analog computer, using squares along with the Darcy equation. Analog techniques are limited to small networks of the order of 100 pipes.

Wood and Charles [90, 1972] describe an interesting digital technique using linear theory. Flow-dependent pipe resistances are iteratively corrected until a balance is achieved. This method requires fewer iterations than Newton-Raphson, which, in turn, requires fewer than Hardy Cross. However, use of linear theory requires that a matrix equal in size to the number of pipes (much greater than the number of loops or nodes) must be inverted.

In reviewing all of the digital techniques, it seems that each one has particular advantages and disadvantages. The fact remains that each technique solves for all heads and/or flows and that the time required to accomplish this does not vary substantially among these more sophisticated methods.

Difficulties in Applying Full Network Balances to Control: For the purpose of control, it is necessary to make simulated projections into the future, using as the forcing function the anticipated time varying demand. A time varying simulation is made up of a series of static solutions. Each uses the demand value appropriate to the time of the solution, as well as current tank depths, calculated from the previous tank depths by numerically integrating the flow into the tanks during the pervious time interval. In order to carry out such a simulation, only heads and flows associated with pumps and tanks need be calculated.

A full network balance requires *excessive computer time* simply be-

[b]Discussions of techniques similar to those mentioned above can be found in bibliographical notes 31, 34, 37, 45, 83, and 92.

cause, at each time interval, it calculates many values not relevant to control. The best balance time for a 250 pipe network is on the order of ten seconds [24]. This is entirely adequate for the purpose of design, but, when contemplating dynamic programming control, which requires thousands of balances, clearly that rate of solution is far too slow.

Any network solution is burdened by the problem of *unknown takeoffs*. All consumptions in the skeletonized network must be assumed to be "taken off" at nodes. In practice, there is never sufficient instrumentation to determine nodal takeoffs, so they must be estimated based upon population patterns. As time passes, the assumed takeoffs must be revised to reflect changing population patterns. In addition, errors are introduced when a full network is approximated by a skeletonized network.

Pipe roughness changes with age, primarily because of the buildup of deposits. For this reason, periodic field measurement programs are necessary to determine the Hazen-Williams C coefficient of major mains. Smaller mains are seldom measured, so guesses at C values are required.

The *network configuration changes* as pipes are added. The complex bookkeeping inherent in all network solution techniques must be revised with each network configuration change.

Empirical Head Drop Expressions

In 1961, M.B. McPherson [51] presented an alternative to the full network solution. A generalized head drop expression calculated the amount by which pump discharge head exceeds tank head in a simple system containing only one tank and one pumping station:

$$\Sigma h/Qd^m = \phi(Qp/Qd)^n, \tag{1.3}$$

where Σh represents head drop from pump to tank, made up of the sum of many individual head drops across the district; Qd, total system demand; and Qp, pump flow. The constants ϕ, m, and n are found from three data points. This expression was derived under the assumption that all demands rise and fall together (proportional loading). Since tank head follows from tank depth, this expression provides pump discharge head, and, hence, pump flow. Owing to the functional form of Equation (1.3), pump flow cannot be solved for directly; instead, it must result from an iterative procedure. Nevertheless, the balance time is orders of magnitude faster than that of a full network solution. Given total demand, tank flow follows from pump flow, thus, a very rapid time simulation is possible by use of numerical integration.

The generalized expression has several serious disadvantages. First, the one-tank, one-pumping-station requirement is quite restrictive since many pressure districts contain more of each. Second, because ϕ and Qp are always positive, it is not possible for Σh to take on a negative value. A negative value of Σh means that tank head is greater than pump head. This can happen when a small pump is put on line after a tank is filled. Finally, the accuracy of the time simulation in duplicating actual operating data was never demonstrated because the systems dealt with were hypothetical extensions of existing systems.

In a later study, McPherson [52] sought to develop an improved generalized expression but decided that the existing one was better than any others proposed. In an appendix to this report, Davis [15] presented two head drop expressions that he found to be less satisfactory than McPherson's. These expressions, attributed to M.H. Diskin and R.W. Adams are, respectively:

$$\Sigma h = [C_1 - C_2(Qd/Qp)]Qp^m, \tag{1.4}$$

$$\Sigma h = K\,Qd + C. \tag{1.5}$$

The constants C_1, C_2, and m in Equation (1.4) and K and C in Equation (1.5) are found from data points. While these and McPherson's expressions have their disadvantages, they serve as a basis for the modeling presented in this work.

Current Methods of Control

In most pressure districts, control is effected by putting discrete pumps on and off line at various pumping stations. By use of various parallel combinations, pumping stations can have as many as ten flow ranges. Each pump combination has a nominal flow capacity, but owing to the variation of suction and discharge head, the flow varies over a range of up to 20 percent of the nominal flow. Less frequently encountered means of control include the continuous adjustment of variable speed pumps or throttling valves. Continuous controls are not considered in this work.

Levels of Control

Manual control is the simplest though most expensive way to control pumps. In this case, all pumping stations are manned, and pump change decisions are largely a function of nearby tank depths. Some degree of coördination can be achieved by telephone conversations among operators.

Most pump station personnel have been eliminated by many cities in the conversion to *supervisory control.* Key system variables, such as tank depths and flows, and pump heads and flows, are telemetered to a central location at which a single operator makes pump change decisions. While telephone links are most frequently used, Philadelphia employs a solid-state microwave link [66]. Frequently, accompanying supervisory control systems are data logging computers that record operating data and alert the operator to abnormal conditions. In some pressure districts, regardless of the level of control, there are limited local automatic controls that cycle pumps on and off to keep an accompanying tank depth within specified limits.

Computer control is in effect in some localities. At this point, in most cases, the computer takes the place of a supervisory control operator. The computer makes pump changes based upon complex logic sequences, usually built up as the composite experience of several operators. In some cases, the logic sequences are designed with the aid of a network balance computer program in an effort to arrive at economical operation. While some authors anticipate the eventual application of various operations research optimization schemes, only very rudimentary ones have been described in the literature thus far [18].

Existing Water Distribution Control Installations.

Table 1-1 shows the level of control achieved by various cities and counties. The table was compiled from the open literature and private conversations with several water department officials from major cities.

Related Studies in Control and Optimization

The original article published concerning closed-loop control of water system operation was written by D. Brock [5] in 1963. A full network solution is used to aid in the control of a two-tank pressure district in Dallas, Texas. Selected flow measurements are used to scale takeoffs. Various pump combinations are periodically simulated to find the most efficient way to force tank depths to follow preestablished profiles. The emphasis is on short-term efficiency, and no rationale for the tank depth profiles is given.

In 1965 and 1966, M.B. McPherson presented a series of articles [53-57] in which he used extended time simulations, made possible by the generalized head drop expression, to investigate various control methods. The pressure districts simulated are basically existing districts, modified in

Table 1-1
Existing Water Distribution Control Installations

	Data telemetry	*Recorder/ Data logger*	*Supervisory control*	*Limited local automatic controls*	*Control studies in progress*	*Computer logic diagram control*	*Supervisory control aided by network solution program*	*Optimization in planning stages*	*Reference*
Birmingham, Ala.	x		x						*67*
Santa Clara County, Calif.	x	x	x						*67*
Winnipeg, Manitoba, Canada	x	x	x	x					*67*
Denver, Colo.	x	x	x		x				*13, 81*
St. Petersburg, Fla.	x	x	x						*67*
Honolulu, Hawaii	x	x		x	x	x	x	x	*2*
Chicago, Ill.	x	x	x	x					*67*
St. Louis Park, Minn.	x		x						*67*
Albuquerque, N.M.	x	x	x	x	x				*67*
Monroe, County, N.Y.	x	x	x			x		x	*22*
Philadelphia, Pa.	x	x	x		x				*1, 66, 69*
Memphis, Tenn.	x		x						*67*
San Antonio, Tex.	x	x	x			x			*82*
Dallas, Tex.	x	x	x	x			x	x	*7, 28*
Houston, Tex.	x	x	x		x	x			*75*
El Paso, Tex.	x	x	x		x			x	*67*
Seattle, Wash.	x	x	x						*67*
Madison, Wisc.	x		x						*67*
Wauskeska, Wisc.	x					x	x		*38*

order to make the generalized expression applicable. The various pumps simulated are nearly the same size. The conclusion is reached that a "floating tank" policy, characterized by few pump changes and large tank depth excursions, is more economical than a policy where tank depths are brought to a maximum nightly. From this study, he concludes that "near-uniform pumping implies efficient operation" [53]. A three-way comparison is made among equalizing storage operation using an elevated storage tank, direct pumping with no storage; and ground level storage, where a booster pump draws water from ground storage part of the day. The conclusion is that there is no substantial pumping cost difference among these configurations, but the cost does depend upon pump scheduling.

In a recent report to the American Water Works Association Research Committee on Distribution Systems [58] McPherson described operational control as follows:

> As generally conceived, under normal operating conditions, this [control] would be accomplished in the following sequence: instantaneous assembly of field intelligence; prediction of the most probable imminent loading; system simulation under that loading using all feasible alternate strategies; selection of the preferred option; and then transmitting orders for effectuating that choice by automatic actuation of field controls. These steps might be cycled as frequently as every quarter hour.

This concept of control will serve as the contrast to the concept developed in the chapters which follow.

In an application of Operations Research techniques, Stephenson [78, 1970] uses linear programming to set daily flows in large transmission mains. Cost minimization due to pumping is accomplished by considering various source water costs and anticipated regional demands. However, the control is open loop and thus is not able to react to change.

Dynamic programming (DP) has been used for the purpose of optimization in the water resource area for about ten years. Buras [9, 1963] and Young [91, 1967] were concerned with the operation of river reservoirs. DP determines optimal sequences of drafts from the reservoirs subject to stochastic inputs. Meier and Beightler [61, 1967] have extended the problem to multiple reservoirs on branching streams. Buras and Schweig [10, 1969] have used DP in aqueduct routing, while Matucha [47, 1972] used DP to find economical operating policies for the large California aqueduct. Similarly, Kalley [36, 1969] used DP to size a single pipeline over varying terrain, while Wong and Larson [88, 1968] were concerned with finding operating policies in a natural gas pipeline. Finally, Schaake and Lai [39, 1970] used both linear programming and DP for a model for capacity expansion planning of water distribution networks. Other references making similar use of optimization are included in the bibliography [6, 11, 16,

29, 34, 71]. While none of this work is directly applicable to the distribution control problem, it provides a good background to related DP applications.

The Systems Approach

As is typically the case in most fields, the majority of technical articles written in the area of water distribution design and control are directed toward quite specific problems. A few authors do address broad distribution problems with a systems approach using current techniques of modeling and optimization. Some notable examples follow.

Existing Systems Work in Design and Control

In 1971, deNeufville *et al.* [19] laid out a general systems approach to the design of water distribution systems. The concern of that paper is to direct designers' efforts toward consideration of purposes, costs, and benefits at a time when computers are lifting computational burdens. Suggested design steps can be summarized as follows:

1. Definition of multiple objectives.
2. Formulation of measures of effectiveness.
3. Generation of design/operation alternatives.
4. Evaluation of alternatives in terms of the multiple objectives and measures of effectiveness.
5. Selection of preferred design by balancing the importance of each objective.

This approach was used to find the proper balance between the size of water transmission tunnels and pumping in New York City in a study by deNeufville [19].

Sturman [79, 80, 1971] has applied a systems approach to the design of a water supply and distribution system to meet requirements over a planning period. The approach here is to some extent, directed more specifically toward water distribution, and it contains the important step of optimizing the selected system. Mathematical programming is used along with the associated decision variables, cost expressions, objective functions, and constraints. Specifically, Sturman draws upon the DP and linear programming work of Schaake and Lai [39].

Neel [64, 1971] presents an excellent background article on the systems approach to computer control of water distribution. The benefits to be enjoyed by many classes of water users are described. Three basic models are used: a use-prediction model, a model of the distribution system, and a model of system constraints. These models are to feed into an optimizer to determine the optimum output schedule.

Much of the work to follow will include the development of the models and an optimizer as described generally by Neel. However, the scope of the problem will be limited by several assumptions.

Objectives of the Study

The purpose of the current work is to develop a technique to control tank-dominated distribution systems, with the following objectives:

1. Delivery of all the water required to customers at adequate pressure.
2. Adequate emergency reaction.
3. Minimization of operating cost without any changes in existing equipment.
4. Holding the number of daily pump combination changes to the approximate current level in order to limit wear.
5. Conforming to all system operating constraints.

The review of water distribution design and control literature has suggested the need for study in several areas to overcome problems in meeting the above objectives. The work has been divided into four main areas, which may be summarized as follows.

Chapter 2 presents a *macroscopic distribution system model,* which is developed somewhat along the lines of McPherson's generalized model [51, 1961]. The model is capable of solving for major pressures and flows in a pressure district containing any number of pumping stations and elevated tanks. The model is characterized by rapid and accurate balances because the solution is noniterative and is built upon existing operating data. The capability of the model to reproduce past operating data by time simulation is demonstrated. A study of transient and steady-state responses of this model helped to increase an understanding of the dynamic nature of a water distribution system.

An *adaptive model* developed in Chapter 3 has the purpose of always keeping the macroscopic system model up to date. The model adapts to changing conditions by means of a Kalman filter, which continuously analyzes the operating data. In this way, changes of piping configuration and parameters in the field do not require computer program changes.

In order to understand the forcing function, a *statistical study of time varying demand* is made in Chapter 4. Periodicities in demand are detected by Fourier analysis. Statistical tests determine the number of significant harmonics, as well as days of the week or portions of the year that are significantly different from one another.

Finally, in Chapter 5, system control is developed by use of dynamic programming incorporated with the fast macroscopic model along with the demand mean-value function. Minimization of pumping energy coupled with maintenance of all tank depths within specified limits serve as the objectives of the controller. The DP method as presented here leads to a time varying subdivision of tank depths into regions thus signifying the choice of pumps to use. Simulations show the superiority of DP control over the often advocated floating tank policy.

Assumptions

1. Pressure districts will consist of pumping stations, pipes, and elevated storage tanks. Continuously adjustable valves or pumps and in-line booster pumps will not be considered.
2. Pressure districts will be primarily residential, and, while the controller will react in such a way to meet emergency conditions, the optimization will apply only to normal operations.
3. No modification to existing equipment will be specified.
4. The cost of water from all sources will be identical.
5. Tank depths that will lead to adequate customer pressure will be assumed to be given as the result of a previous design study.
6. A final assumption will be that an electric power demand limit exists under which the pumps available for operation are given, and under which operating cost is proportional to energy consumption. Such a power demand limit specifies the maximum power in kilowatts to be drawn at a pumping station. While power consumption varies slightly as a pump meets various head conditions, large step changes result from pump changes. A specific demand limit has the effect, under the threat of a penalty charge, of making larger pumps unavailable for operation.

The above assumptions are made largely to limit the problem. With additional attention to detail, each could be eliminated with the exception of the first. The inclusion of continuous control elements would considerably alter the problem and would require a separate control study. However, as a practical matter, very little continuously variable equipment is currently used in the water distribution industry.

Specific System to be Studied

There is no truly typical water distribution system or pressure district;

therefore, water distribution studies must in actuality be reduced to studies of various specific examples. Thus, while all modeling and control work is carried out in general, all specific examples will be related to the Torresdale High Service-Fox Chase Booster district in Philadelphia. This district, which contains two pumping stations and two elevated storage tanks, is among the more complex districts in the city.

2 Distribution System Modeling

Introduction

The primary purpose of this chapter is to present a concept in distribution system modeling developed especially for the purpose of control. Model accuracy is established by a time simulation validation procedure that can be applied equally well to any modeling procedure. Finally, a two-pumping-station, two-tank pressure district is modeled, and the corresponding steady-state and transient responses are examined.

Model Requirements for the Purpose of Control

Regardless of the way in which distribution control is effected, the capability of fast time simulation is desirable. In looking toward dynamic programming, speed is essential, although speed is of little value without accuracy. The problems of unknown takeoffs inherent in a network model must be overcome. Finally, the model must adapt readily to ever-present system changes.

The model derived here might be described as a macroscopic model because it calculates only major pressures and flows, as opposed to a microscopic network model, which calculates all flows and pressures. For the purpose of control, only the conditions surrounding pumps and tanks need be known, so frequently less than 1 percent of the values calculated by a network balance are required to carry out a time simulation needed for this purpose. It is reasonable to suppose that a model that calculates many times fewer values than a network model should be capable of carrying out the calculation much faster.

Macroscopic Variables

What might be called macroscopic variables, necessary for a time simulation, are those pressures (or heads) and flows associated with pumps and tanks. Pump performance is characterized by its suction pressure, discharge pressure, and flow. A tank is characterized by flow and depth change. While taking no direct part in a dynamic simulation, internal pressure points provide an indication of pressures available to customers.

A macroscopic model consists of empirical expressions relating macroscopic variables. These empirical relations are obtained from analysis of past operating data. Typically, these expressions are much fewer in number than the set of equations describing the detailed nonlinear network. Speed is attained because few variables are calculated, and accuracy results from the fact that the model remains current, since it is built from current data. It will be shown that in comparison to a full network model under strictly proportional loading, the macroscopic model accuracy can be esentially perfect; but that accuracy, while still very good, deteriorates somewhat as real data are used where the loading is less than exactly proportional. For this reason, the model is primarily applicable to largely residential districts not subject to large nonproportional industrial loads. While the macroscopic model is not capable of handling large nonproportional loads, full network models also have difficulties in this respect.

While the macroscopic model is driven by total demand only, a full network model is driven by many nodal demands, usually assumed to vary proportionally. A network model would be presumably capable of replicating an industrial district if the nonproportional loads could be assigned to the correct nodes at the correct times. This is a particularly difficult "if," especially when projecting into the future. Since, unfortunately, the study of different districts is reduced to the study of special cases, it will have to be determined by experience at what point, as districts become more industrial, a network model accuracy exceeds the macroscopic model accuracy, if, indeed, it does.

A final word of caution is necessary in applying the macroscopic model. Since the model is derived from operating data, it is possible to model a system other than that envisioned by the designers. There are frequently many unmonitored valves within a pressure district that may be inadvertently left at other than their nominal positions. For this reason, it has been strongly recommended [59] that prior to macroscopic modeling, a complete survey of all valves be made. A full network balance solution can be helpful in detecting erroneous valve settings. The macroscopic model will model a district exactly as it sees it, so modeling accuracy will not suffer from erroneous valve settings. However, since districts are presumably designed for efficient operation with a given nominal configuration, there will be a potential for more efficient operation if the nominal configuration is attained before control studies are made.

Time Simulation

A single system balance, whether it is carried out by a network model or the macroscopic model, is a static simulation. To make a dynamic simulation as shown in Figure 2-1, tank depth is calculated as the integral of the flow into a tank. Since a full network balance is an iterative procedure, a closed-form solution for tank depth as a function of time is obviously impossible. Likewise, it will be shown that a closed-form solution is impossible for the macroscopic model, so, in either case, numerical integration is necessary. Figure 2-1 shows the basic order of computation needed to carry out a dynamic time simulation.

Pump and Tank Equations

Regardless of the way a system balance is carried out, pumps must be described by a head-flow relationship. A typical pump curve is shown in Figure 2-2. A frequently used equation relating pump increase, Hp, to pump flow, Qp, is as follows:

$$Hp/Hpd = C + (1-C)(Qp/Qpd)^2. \tag{2.1}$$

A design point is usually specified by Hpd and Qpd, the pump design head and flow. The zero flow condition is characterized by the pump shutoff head, $Hps = C\ Hpd$. The constant C has a value greater than one and usually less than two.

Some authors prefer to express pump head in terms of three or four powers of pump flow, but since the operating range of a pump is quite narrow, the additional terms make little difference. Experience has shown that far more important than using an elegant pump equation is keeping a simple one up to date. Pump head-flow characteristics change as pumps age and for other reasons that are difficult to determine [69]. In this study, for example, seasonal variations in pump characteristics lead to simulation inaccuracies until the problem was diagnosed.

Elevated storage tanks are characterized by a relationship between storage volume and water depth. In most cases, the normal operating range of tank depth is within the cylindrical portion, so there is a constant relation between depth and volume. In some cases, the shape of the tank causes the relation to be a function of tank depth. If tank flow data collected by a data logger are averaged over the collection period, there should be a straight line relationship between tank flow and change in tank depth. When periodically printed flow values are instantaneous values rather than averages, a considerable point scatter can result when plotting change in tank depth against tank flow. In this case, the tank characteristic can be found by passing a regression line through the points.

The relation between tank depth and flow is as follows:

$$Dt = Kt \int_{T_0}^{T} Qt\ d\xi + Dt(T_0) \tag{2.2}$$

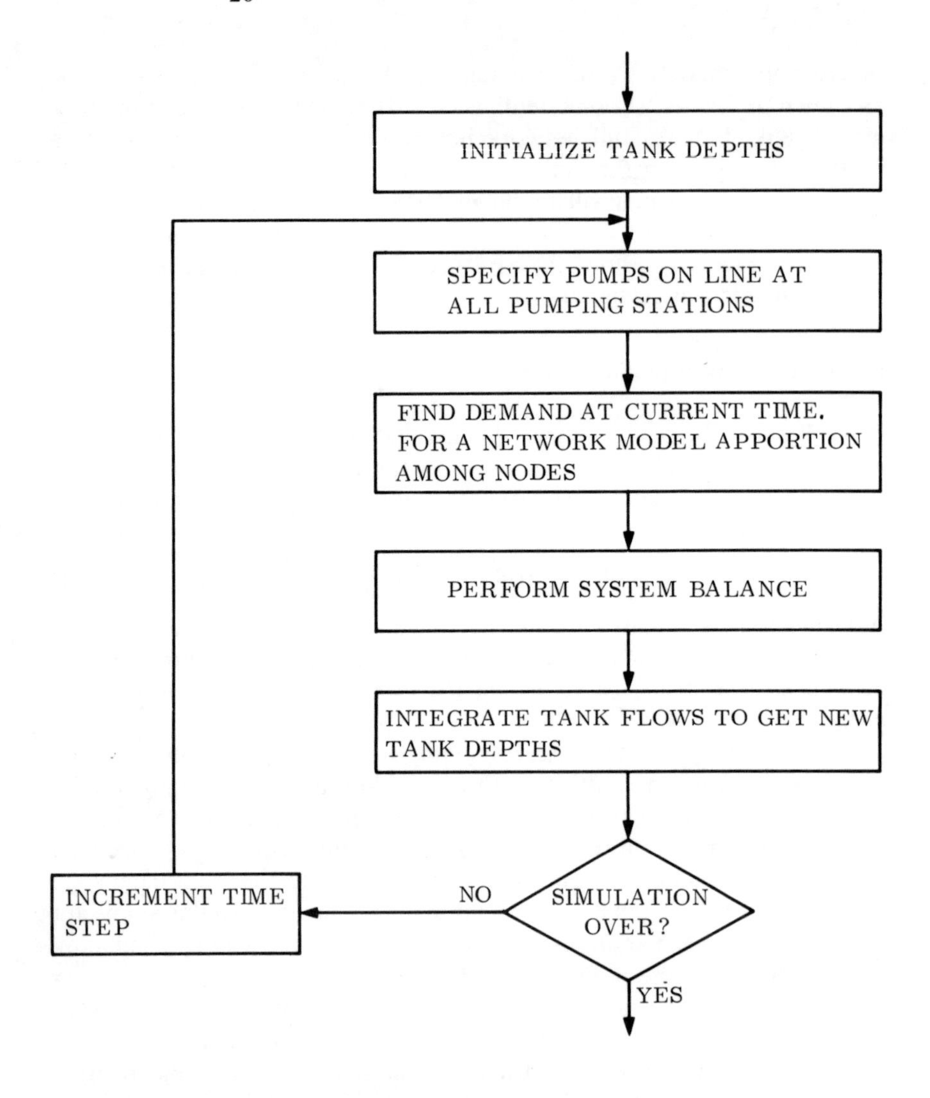

Figure 2-1. Basic Pressure District Dynamic Simulation

where *Dt* is the tank depth (in feet); *Qt*, tank flow (in million gallons/day); *Kt*, tank characteristic (in feet/million gallons); and ξ is time (in days). It is possible to represent hydraulic elements by equivalent electric circuits (see Figure 2-3). Head, or depth, being a potential, is analagous to voltage and will be represented by an open arrow (→). Flow and current are analagous and will be represented by a closed arrow (➝). The equivalent circuit of a tank is a capacitor, as shown in Figure 2-3.

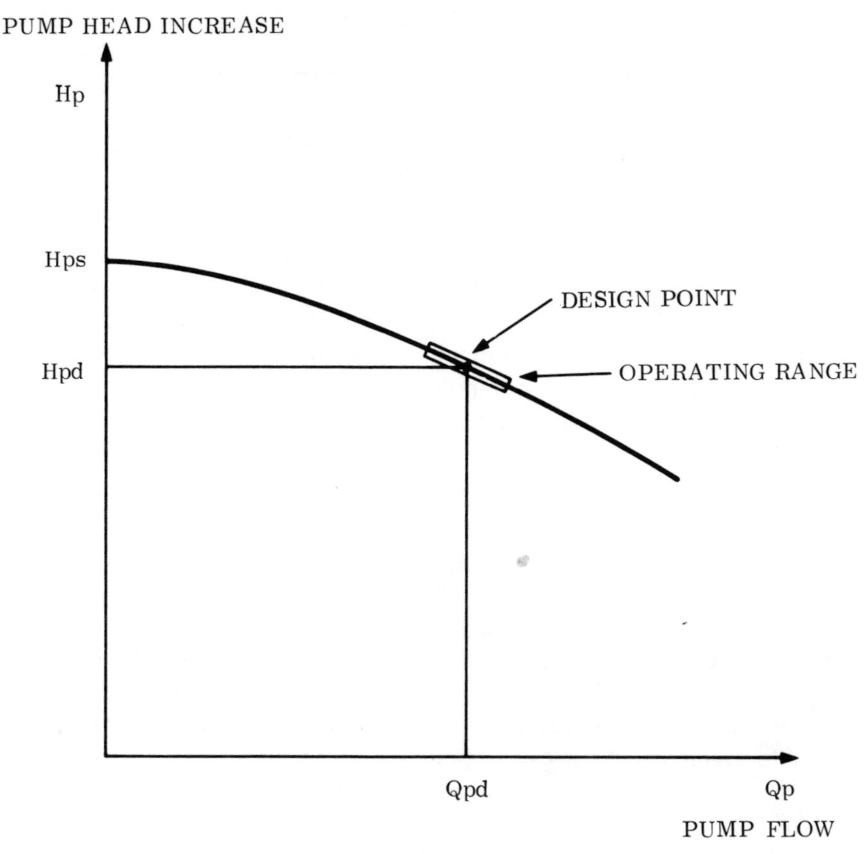

Figure 2-2. Typical Pump Curve

The Macroscopic Model

The purpose of the macroscopic model is to perform a system balance. Given tank depths, pumps on line at each station, and total demand, the model calculates pump discharge head and flow, tank flow, and selected internal pressure points. The model consists of a set of equations derived from both conventional pump and tank equations and from empirical equations developed for this purpose.

Types of Empirical Equations

There are three major types of empirical equations, which are summarized as follows:

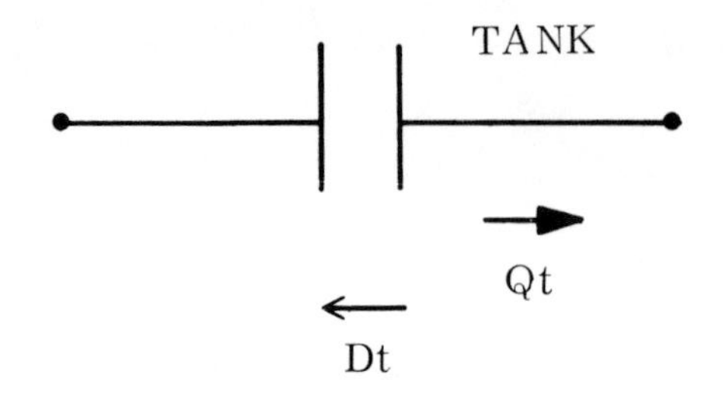

Figure 2-3. Tank Equivalent Circuit

I. Head drop from each pumping station to each tank in terms of total demand flow and all pump flows.
II. Tank flow in terms of total demand flow, all tank heads, and all pump flows.
III. Internal pressure points in terms of total demand flow, all pump discharge heads, and all tank heads.

Given tank depths (heads), equation type I fixes pump discharge pressure and, therefore, pump flow. Pump flow is also a function of pump suction head and the pump combination on line. Equation type II provides the tank flow necessary for dynamic simulation by numerical integration. Equation type III does not play an active part in the dynamic simulation, but the calculated pressure serves as an indication of model accuracy and can be used to monitor weak pressure points in the system.

Unknown coefficients in the empirical expressions are calculated by use of multiple regression analysis of operating data. Redundant empirical expressions and redundant independent variables within the empirical expressions are eliminated by use of statistical tests associated with multiple regression.[a]

The general pressure district to be modeled by the macroscopic model contains I pumping stations, J elevated storage tanks, and H internal pressure points. Indices referring to these elements are, respectively, i, j, and h. At any time, pump combination $k(i)$ is on line at station i. Table 2-1 summarizes this nomenclature.

Pump Related Equations

Pumping Station Head-Flow Relation:

$$Hp(i) = Cp(i, k, 1) + Cp(i, k, 2)Qp(i)^{1.85}, \tag{2.3}$$

[a]See Appendix A or bibliographical note 32 for a description of multiple and stepwise regression.

Table 2-1
General Pressure District Nomenclature

Index	*Component*
$1 \leqq i \leqq I$	Pumping stations
$1 \leqq j \leqq J$	Elevated storage tanks
$1 \leqq h \leqq H$	Internal pressure points
$1 \leqq k(i) \leqq K(i)$	Pump combination number in operation at station i

where $Hp(i)$ represents head increase across station i; $Qp(i)$, pump flow at station i; and Cp, pump constant array.

As the arguments indicate, a different set of constants, Cp, applies to each pump combination at each station, The constants Cp should be determined by regression analysis of operating data rather than from manufacturers' curves. Equation (2.3) applies to the station characteristics, rather than individual pump characteristics, owing to station pipe losses. Pump head is related to flow raised to the exponent 1.85 rather than flow raised to the exponent 2.0 as shown in Equation (2.1). As already pointed out, the difference is negligible as long as the constants Cp are correctly determined. The purpose in using the 1.85 exponent is to permit a noniterative solution for pump flow.

The electrical equivalent circuit element of a pumping station is a flow-dependent head (potential) source is shown in Figure 2-4.

Pump Suction Head:

$$Hs(i) = Cs(i, 1) + Cs(i, 2)Qd^{1.85} + Cs(i, 3)Qp(i)^{1.85}, \qquad (2.4)$$

where $Hs(i)$ represents suction head at station i; Qd, total demand flow; and Cs, suction head constant array. Suction head is often esentially constant, but, depending upon the water source, it may vary as a function of total demand and/or pump flow. The constants Cs are determined by multiple regression, with statistically insignificant ones eliminated by stepwise regression. Note that head is related to flow raised to the exponent 1.85. This exponent is used because head variations throughout the system are a function of resistive losses, as described by the Hazen-Williams equation [Equation (1.2)]. Although the suction head equation is an empirical relation, it is not classified along with equation types I, II, and III because of its minor importance. If suction heads are not constants, they usually vary little. The electrical equivalent circuit element of suction head is a pump-flow-dependent and demand-flow-dependent head source, as shown in Figure 2-5.

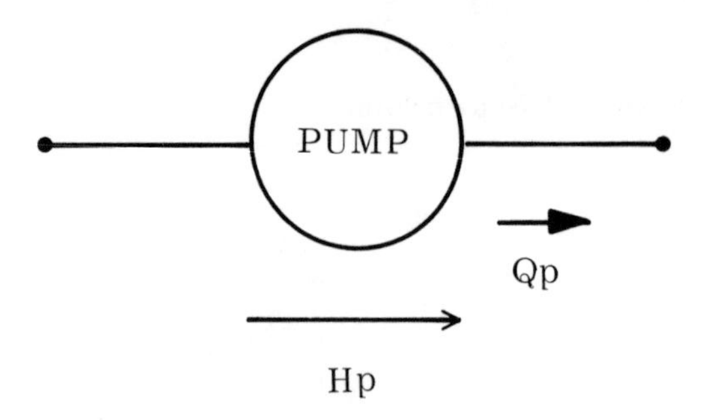

Figure 2-4. Pump Equivalent Circuit

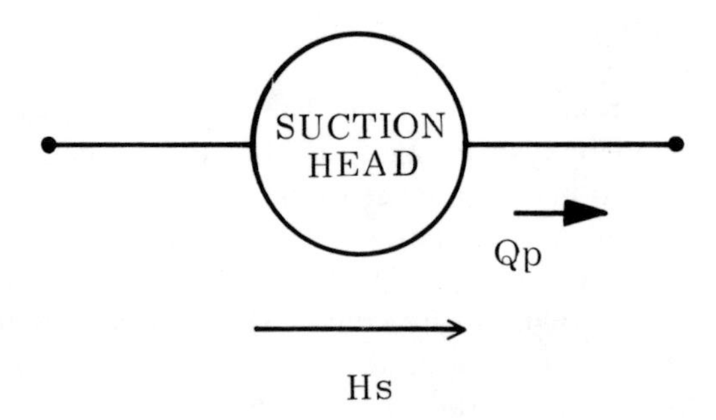

Figure 2-5. Suction Head Equivalent Circuit

Pump Discharge Head:

$$Hd(i) = Hs(i) + Hp(i), \tag{2.5}$$

where $Hd(i)$ represents discharge head at station i; $Hs(i)$, the suction head at station i; and $Hp(i)$, the station i head increase. Pump discharge is specified relative to a given datum. The equivalent circuit is a combination of those for suction head and pump head, as shown in Figure 2-6.

Pump to Tank Head Drop (Equation Type I)

The pump to tank head drop expression, designated as empirical expression type I, is analagous to McPherson's [51] generalized head drop expression:

$$\begin{aligned} Hpt(i) = {} & Cpt(i, 1) + Cpt(i, 2)Qd^{1.85} \\ & + \sum_{m=1}^{I} Cpt(i, m+2)Qp(m)^{1.85}, \end{aligned} \tag{2.6}$$

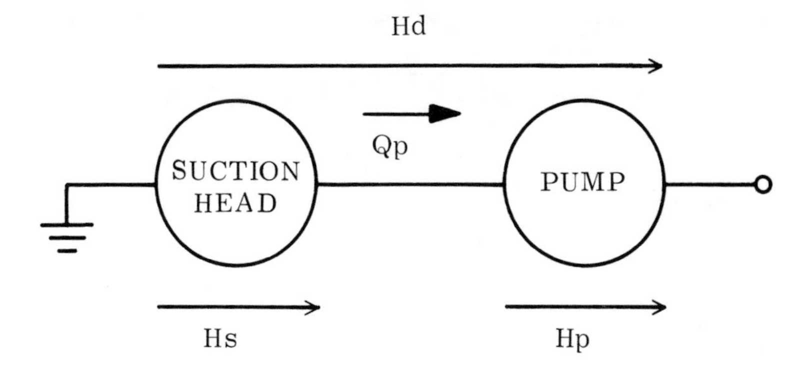

Figure 2-6. Pump Discharge Head Equivalent Circuit

where $Hpt(i)$ is the head drop from pumping station i to its reference tank, and Cpt represents constant array to be determined by multiple regression. The head drop from pumping station to tank is a potential drop dependent on demand flow and possibly on all pump flows (see Figure 2-7). Refer to Figure 2-8 for three examples of head drop from pump to tank (pump discharge head minus tank head). The hydraulic grade line connecting pump discharge head to tank head is a hypothetical head profile across the distance separating pump and tank. In case A, a large pump is on line at a time of low demand, so all heads across the pressure district are greater than the tank head, causing the tank to fill. Suppose that, in case B, the same pump is on line, tank depth is the same, but demand is greater. In this case, some district heads are less than the tank head, so while pump discharge head is still greater than tank head, the tank is draining. Note that since demand is greater in this case, head drop from pump to tank has decreased, causing the pump to operate at a greater flow with reduced discharge head. Finally, in case C, a small pump is on line when demand is high, so pump discharge head is actually less than tank head, causing a negative head drop from pump to tank. In Equation (2.6), the coefficients of pumping station flows, $Qp(i)^{1.85}$, are generally positive because greater flow into the system causes greater resistive head drop. On the other hand, the coefficient of $Qd^{1.85}$ is negative, permitting the transition from case A to case B in Figure 2-8, and, ultimately, the negative head drop in case C.

Initially, $I \cdot J$ regressions are carried out to find $Hpt(i,j)$ for $1 \leqq i \leqq I$ and $1 \leqq j \leqq J$. Only one head drop expression is needed for each pump station i, so the redundant ones are eliminated by keeping only that expression which has the greatest multiple correlation coefficient after the regressions are carried out. This process results in a reference tank for each pumping station, identified by $Tr(i)$. This means that the discharge head at station i is found by adding the head drop $Hpt(i)$ to the head of tank $Tr(i)$. Pumps and

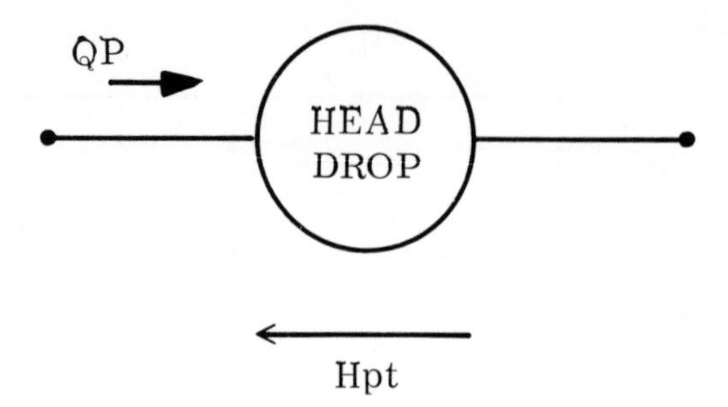

Figure 2-7. Network Head Drop Equivalent Circuit

tanks that are close to one another will usually be paired off, but this must be ascertained, as described, from the operating data rather than assumed. It is possible for a tank to act as a head reference to more than one pumping station. Finally, statistically insignificant independent variables are removed by stepwise regression.

Tank Equations

Tank Head:

$$Ht(j) = He(j) + Dt(j), \tag{2.7}$$

where $Ht(j)$ represents the head of tank j; $He(j)$, the tank bottom elevation with respect to a common datum; and $Dt(j)$, the tank depth. Tank head is relative to the same datum as pump discharge and suction head. The equivalent circuit for tank head follows from the tank depth circuit (see Figure 2-9).

If, in a particular distribution system, it is likely that a tank depth will reach maximum or minimum limits, provisions are usually made to valve off the tank. When such provisions are made, head drop expressions will have to be maintained between a pumping station and more than one reference tank in case the primary reference tank should be valved off.

Tank Flow (Equation Type II):

$$Qt(j) = Ct(j, 1) + Ct(j, 2)Qd + \sum_{m=1}^{J} Ct(j, m+2)Ht(m)^{1/1.85}$$

$$+ \sum_{i=1}^{I} Ct(j, i+J+2)\,Qp(i)\,, \tag{2.8}$$

where $Qt(j)$ represents flow into tank j and Ct represents constants to be determined by multiple regression. The equivalent circuit element of the tank flow is a current (flow) source (see Figure 2-10). Note that in keeping

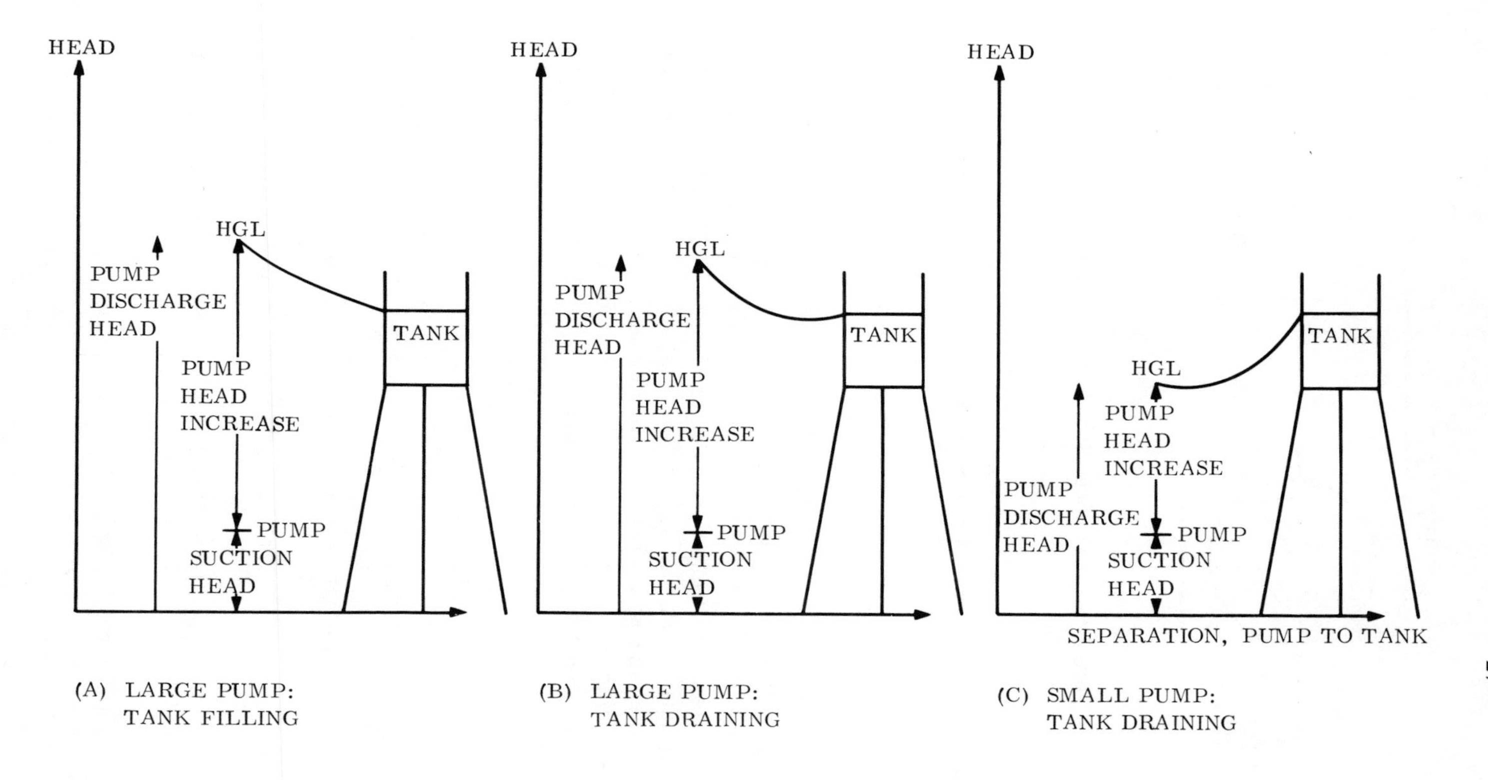

Figure 2-8. Hydraulic Grade Line (HGL) Under Various Loading Conditions

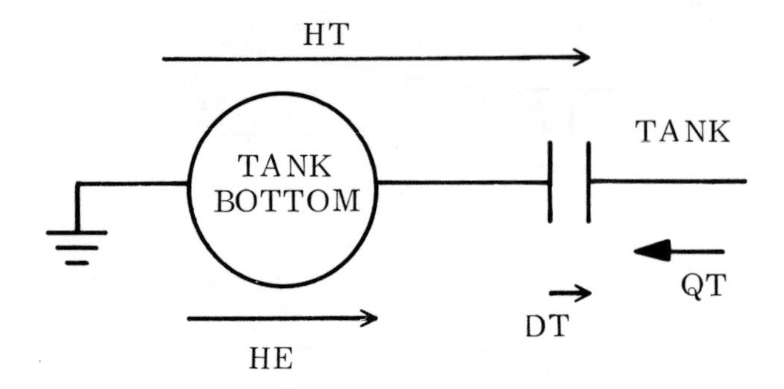

Figure 2-9. Tank Head Equivalent Circuit

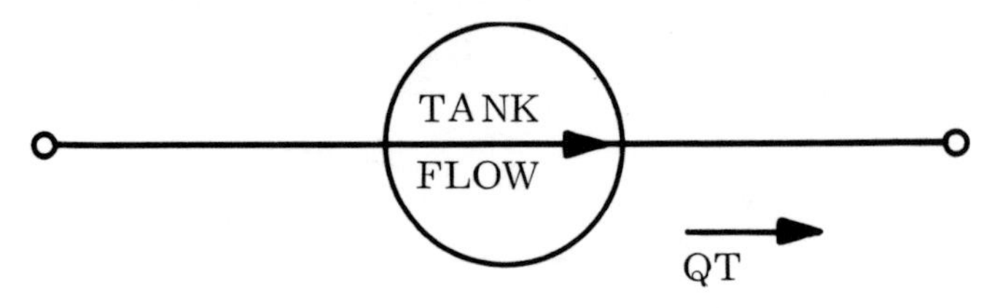

Figure 2-10. Tank Flow Equivalent Circuit

with the Hazen-Williams relation, flow is related to head to the exponent 1/1.85. It is only necessary to have $(J-1)$ tank flow expressions because, given total demand, all pump flows, and $(J-1)$ tank flows, the remaining tank flow results from continuity of flow. Initially, all J regressions are carried out; then the poorest regression, indicated by the least multiple correlation coefficient, is thrown out. As before, insignificant dependent variables are removed by use of stepwise regression. Suppose that the tank whose regression was the poorest is tank number $j = \hat{j}$; then the corresponding constants follow from flow continuity:

$$Ct(\hat{j}, 1) = -\sum_{j \neq \hat{j}} Ct(j, 1), \tag{2.9}$$

$$Ct(\hat{j}, 2) = -\sum_{j \neq \hat{j}} Ct(j, 2) - 1, \tag{2.10}$$

$$Ct(\hat{j}, m+2) = -\sum_{j \neq \hat{j}} Ct(j, m+2), \quad 1 \leqq m \leqq J, \tag{2.11}$$

$$Ct(\hat{j}, i+J+2) = -\sum_{j \neq \hat{j}} Ct(j, i+J+2) + 1, \quad 1 \leqq i \leqq I. \tag{2.12}$$

Tank Depth Rate: Tank depth rate, the time derivative of tank depth, is given as follows:

$$D^{\cdot}t(j) = Cdr(j)Qt(j), \tag{2.13}$$

where $D^{\cdot}t(j)$ is the rate of increase of depth in tank j, and $Cdr(j)$ represents tank depth/volume characteristic previously described.

Internal Pressure Points (Equation Type III)

Selected internal pressure points are a function of total demand and the heads of pumps and tanks:

$$Pn(h) = Cn(h, 1) + Cn(h, 2)Qd^{1.85} + \sum_{i=1}^{I} Cn(h, i+2)Hd(i)$$

$$+ \sum_{j=1}^{J} Cn(h, j+I+2)Ht(j), \qquad (2.14)$$

where $Pn(h)$ is the pressure at internal node h and Cn represents constants to be determined by multiple regression. Again, insignificant independent variables are removed by stepwise regression.

Dynamic Equations

Having presented the empirical equations relating macroscopic variables, we can combine these equations with the pump and tank relations to produce the pressure district dynamic equations. The first step will be to solve for pump flow raised to the exponent 1.85. Given this pump flow, tank flow can then be calculated. Finally, pump and tank heads permit the internal pressure point evaluations.

We have now accounted for all of the hydraulic elements. The equivalent circuit of a single pumping station-reference tank combination is shown in Figure 2-11. The equivalent circuit of the entire distribution system consists of as many of these single pump-tank combinations as there are pumping stations. The sum of all terms $(Qp - Qt)$ is then the total demand flow, Qd.

$Ht(j)$, $He(j)$, and $Dt(j)$ are the head, bottom elevation, and depth of tank j, respectively. Each pumping station, i, has a reference tank $Tr(i)$. The corresponding head, bottom elevation, and depth are $Hrt(i)$ $Hre(i)$, and $Drt(i)$, respectively. In order to solve for pump flow, raised to the exponent 1.85, pump discharge head (suction head plus head increase) is set equal to the head of the reference tank plus head drop from the pump to that tank (see Figure 2-11):

$$Hd(i) = Hrt(i) + Hpt(i). \qquad (2.15)$$

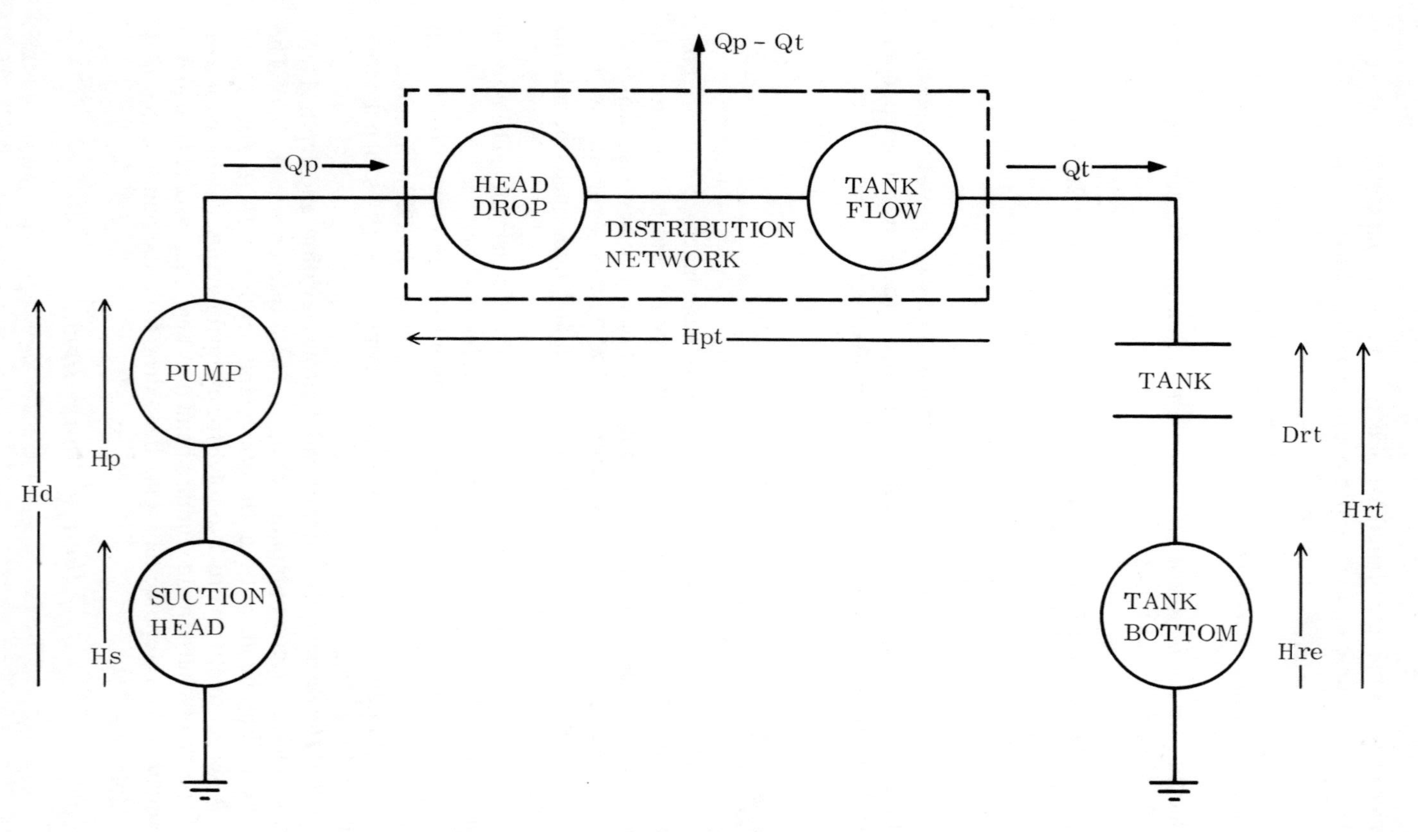

Figure 2-11. Single Pump/Tank Equivalent Circuit

By referring to Equations (2.3) through (2.7), we can expand this expression as follows:

$$Cs(i,1) + Cs(i,2)Qd^{1.85} + Cs(i,3)Qp(i)^{1.85} + Cp(i,k,1) + Cp(i,k,2)Qp(i)^{1.85}$$
$$= Drt(i) + Hre(i) + Cpt(i,1) + Cpt(i,2)Qd^{1.85}$$
$$+ \sum_{m=1}^{I} Cpt(i, m+2)Qp(m)^{1.85}. \tag{2.16}$$

Given I of these equations, it is clear that the I unknown values $Qp(i)^{1.85}$ can be solved for as a function of total demand Qd, tank depths $Drt(i)$, and the pumps in operation $k(i)$.

Simplification of Notation. In order to simplify the notation, a few matrices and vectors will be defined. The general matrix definition is

$$\mathbf{M} = [M(r, c)]_{(R\times C)},$$

where there are $1 \leqq r \leqq R$ rows and $1 \leqq c \leqq C$ columns.

Constants: $\mathbf{C}_1$, $\mathbf{C}_2$, $\mathbf{C}_3$, and $\mathbf{C}_4$ are defined as

$$\mathbf{C}_1 = [Cpt(r, c+2)]_{(I\times I)}, \tag{2.17}$$

$$\mathbf{C}_2 = \text{DIAG}[Cs(r, 3) + Cp(r, k, 2)]_{(I\times I)}, \tag{2.18}$$

$$\mathbf{C}_3 = [Cpt(r, 1) - Cs(r, 1) - Cp(r, k, 1) + Hre(r)]_{(I\times 1)}, \tag{2.19}$$

$$\mathbf{C}_4 = [Cpt(r, 2) - Cs(r, 2)]_{(I\times 1)}. \tag{2.20}$$

Pump Flow: $\mathbf{Qp}$ is defined as

$$\mathbf{Qp} = [Qp(r)]_{(I\times 1)}. \tag{2.21}$$

Pump Flow Raised to Exponent 1.85: $\mathbf{Qp}^{1.85}$ is defined as

$$\mathbf{Qp}^{1.85} = [Qp(r)^{1.85}]_{(I\times 1)}. \tag{2.22}$$

Reference Tank Depth: $\mathbf{Drt}$ is defined as

$$\mathbf{Drt} = [Drt(r)]_{(I\times 1)}. \tag{2.23}$$

By rearranging Equation (2.16) and by applying the above matrices and vectors, we obtain the station pump flow solution:

$$\mathbf{Qp}^{1.85} = [\mathbf{C}_2 - \mathbf{C}_1]^{-1}[\mathbf{C}_4 Qd^{1.85} + \mathbf{Drt} + \mathbf{C}_3]. \tag{2.24}$$

The matrix to be inverted contains no variables, but it does contain pump constants, so if it is to be evaluated in advance of a simulation, it must be evaluated for all possible pump combinations across all pumping stations.

In order to simplify the tank depth expression, the following matrices and vectors are defined:

Constants: **Cdr** and **Ct** are defined as follows:

$$\mathbf{Cdr} = \text{DIAG}[Cdr(r)]_{(J\times J)}, \tag{2.25}$$

$$\mathbf{Ct} = [Ct(r,\, c)]_{[J\times(2+I+J)]}. \tag{2.26}$$

Tank Head Raised to the Exponent 1/1.85: $\mathbf{Ht}^{1/1.85}$ is defined as

$$\mathbf{Ht}^{1/1.85} = [Ht(r)^{1/1.85}]_{(J\times 1)}. \tag{2.27}$$

Tank Depth Rate: $\mathbf{D^{\cdot}t}$ is defined as

$$\mathbf{D^{\cdot}t} = [D^{\cdot}\mathbf{t}(r)]_{(J\times 1)}. \tag{2.28}$$

Tank depth rate follows from Equations (2.8) and (2.13):

$$\mathbf{D^{\cdot}t} = \mathbf{Cdr}\ \mathbf{Ct} \begin{bmatrix} 1 \\ Qd \\ \mathbf{Ht}^{1/1.85} \\ \mathbf{Qp} \end{bmatrix}. \tag{2.29}$$

The following are defined to simplify the internal pressure expression:

Constant: **Cn** is defined as follows:

$$\mathbf{Cn} = [Cn(r,\, c)]_{[H\times(2+I+J)]}. \tag{2.30}$$

Pump Discharge Head: **Hd** is defined as

$$\mathbf{Hd} = [HD(r)]_{(I\times 1)}. \tag{2.31}$$

Tank Head: **Ht** is defined as

$$\mathbf{Ht} = [Ht(r)]_{(J\times 1)}. \tag{2.32}$$

Internal Nodal Pressure Points: **Pn** is defined as

$$\mathbf{Pn} = [Pn(r)]_{(H\times 1)}. \tag{2.33}$$

From Equation (2.14), the internal pressure points expression is as follows:

$$\mathbf{Pn} = \mathbf{Cn} \begin{bmatrix} 1 \\ Qd^{1.85} \\ \mathbf{Hd} \\ \mathbf{Ht} \end{bmatrix} \tag{2.34}$$

Summary

To summarize the macroscopic model, the dynamic equations [Equations (2.24) and (2.29)] are repeated below and a corresponding block diagram is shown in Figure 2-12.

$$\mathbf{Qp}^{1.85} = [\mathbf{C}_2 - \mathbf{C}_1]^{-1}[\mathbf{C}_4 Qd^{1.85} + \mathbf{Drt} + \mathbf{C}_3];$$

$$\mathbf{D{\cdot}t} = \mathbf{Cdr\ Ct} \begin{bmatrix} 1 \\ Qd \\ \mathbf{Ht}^{1/1.85} \\ \mathbf{Qp} \end{bmatrix}.$$

The Hazen-Williams equation shows the experimental results demonstrate that pipe head drop is proportional to flow raised to the exponent 1.85. For this reason, in the empirical expressions, when head is the dependent variable, flows that appear as independent variables are raised to the exponent 1.85. Likewise, when flow is the dependent variable, heads—as independent variables—are raised to the exponent (1/1.85). As the result of following these principles, in Equation (2.29), tank depth rate—or equivalently tank head rate—is related to tank heads raised to the exponent 1.85. Therefore, a closed-form expression for tank depth as a function of time is impossible. A single network balance consists of the evaluation of Equations (2.24), (2.29), and (2.34), in that order.

Pump Energy Calculation

An auxilliary calculation common to a full network model or the macroscopic model is the calculation of pump energy consumption. Average energy consumption per unit flow pumped is an important measure of the efficiency of a distribution control system. The energy consumed by station i, given by $Ep(i)$, is as follows:

$$Ep(i) = Cpe \int \frac{Qp(i)Hpu(i)dt}{Fpe(i,\ k)}, \tag{2.35}$$

where $Ep(i)$ represents energy (in kilowatt-hours); $Qp(i)$, station pump flow (in million gallons/day); $Hpu(i)$, pump head increase (in feet); $Fpe(i, k)$, efficiency of pump k operation at station i (expressed as a percentage); Cpe = 314 kilowatt-hours/million gallons × feet; and dt is time measured in days. $Hpu(i)$ is the head increase across the pump, which is greater than the head increase across the pumping station, $Hp(i)$, by a quantity equal to the station resistive loss:

$$Hpu(i) = Hp(i) + Rpu(i,\ k)Qp(i)^{1.85}. \tag{2.36}$$

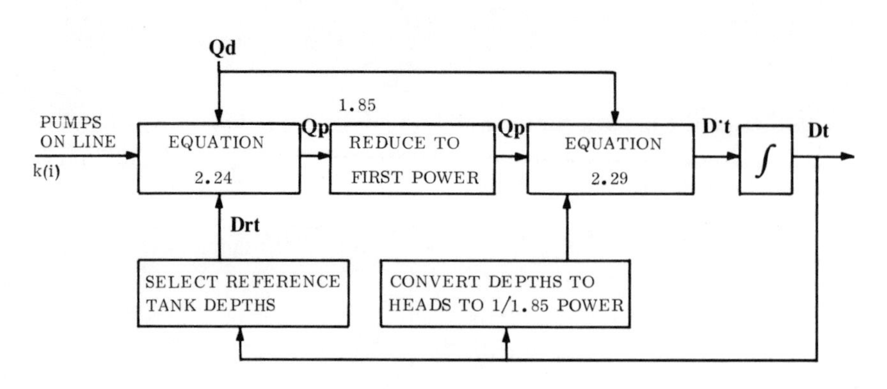

Figure 2-12. Macroscopic Model Dynamics

$Rpu(i, k)$ is the array of station resistance, which varies as a function of the pump in operation, k as a result of different piping configurations. Station resistance follows from a field measurement program.

Pump efficiency may be expressed by a variety of functions. For the present purpose, the following polynomial is used:

$$Fpe(i, k) = \sum_{m=1}^{4} Cpe(i, k, m) \, [Qp(i)/Qpd(i, k)]^{(m-1)} . \qquad (2.37)$$

It is useful to write the expression in terms of the ratio of pump flow to pump design flow so that the same shaped curve can be applied to pumps of varied size. The constants Cpe can be found by polynomial curve fits.

Macroscopic Model Construction and Validation

Summary of Macroscopic Model Construction

The macroscopic model is developed either from actual operating data or simulated operating data produced by a dynamic simulation using a network model, as shown generally in Figure 2-1. A first purpose for constructing a macroscopic model from a network model simulation is to show how very nearly equivalent the two models can be, under proportional loading. Secondly, a new pressure district that exists only on paper can be converted from a network model to a macroscopic model for the purpose of control studies.

The following measurements must be made at periodic intervals throughout the day:

1. Pump station suction head (or pressure).
2. Pump station discharge head (or pressure).

3. Pump flow.
4. Tank flow.
5. Tank depth.
6. Internal pressure.
7. Total demand.

Initially, measurements 1, 2, and 3 are used to develop pump combination equations according to Equation (2.3). Comparison of pump pressure measurements made by gauges mounted physically near individual pumps and station suction and discharge pressures leads to station resistance needed in Equation (2.36). Measurements 4 and 5 are used to develop tank depth/volume characteristics.

The empirical expressions for suction head, head drop from pump to head, tank flow, and internal pressure may be developed from the above measures as follows:

1. Suction head [Equation (2.4)]: measurements 1, 3, and 7.
2. Head drop from pump to tank [Equation (2.6)]: measurements 2, 3, 5, and 7.
3. Tank flow [Equation (2.8)]: measurements 3, 4, and 5.
4. Internal Pressure [Equation (2.14)]: measurements 2, 5, 6, and 7.

In developing the empirical expressions, there are the following points to be observed:

1. Only one expression for head drop from pump to tank is needed for each pumping station. The one expression having the greatest multiple correlation coefficient is retained and the reference tank number is noted.
2. The tank flow expression having the poorest multiple correlation coefficient is removed. Its constants are derived from the remaining expressions through flow continuity.
3. In all expressions, insignificant independent variables are removed by stepwise regression.

Model Validation

Customarily, a distribution system model is said to replicate the system adequately if simulated pressures and flows come close to measured values at a few points in time. This is not the best procedure because random demand variations can make the model appear worse or better than it is. It is preferable to validate a model by performing a time simulation that requires many balances, thus averaging out random fluctuations. The integration of tank flow to obtain tank depth will uncover flow bias errors that can cause depths to drift too high or too low in time.

In order to validate a model, a set of operating data is required, consisting of N time points of total demand, pump discharge heads and flows, tank depths and flows, and internal measured pressures. The demand serves as a forcing function, and the pump combinations in operation at each station can be inferred from station flow. The quality of the model can be judged by the accuracy with which the model simulates actual data. Measures of accuracy for each measurement include error average and error standard deviation. In order to assign a single figure of merit to a model validation, a modified multiple correlation coefficient is used.

Suppose that $D(n, m)$ is the reference array, where there are $1 \leqq n \leqq N$ time points and $1 \leqq m \leqq M$ measurements. The simulated values produced by the model (macroscopic, network, or any other) are $\hat{D}(n, m)$, the difference between simulated and actual values are the errors:

$$E(n, m) = \hat{D}(n, m) - D(n, m). \tag{2.38}$$

Error average for measurement (m) is:

$$\bar{E}(m) = \frac{1}{N} \sum_{n=1}^{N} E(n, m). \tag{2.39}$$

Error standard deviation is

$$\sigma_{E(m)} = \left\{ \frac{1}{N-1} \sum_{n=1}^{N} [E(n, m) - \bar{E}(m)]^2 \right\}^{1/2}. \tag{2.40}$$

The multiple correlation coefficient is normally defined as follows:

$$\rho = \left(\frac{SST - SSE}{SST} \right)^{1/2}. \tag{2.41}$$

SST is the total sum square deviation of all measured values, $D(n, m)$, from their respective averages. SSE is the sum square of the error terms, $E(n, m)$. As a model error approaches zero, ρ approaches unity. All measured values are not of the same units, so in order to cause all measurements to contribute equally to the multiple correlation coefficient, each measurement will be normalized by dividing it by its appropriate standard deviation.

If

$$\bar{D}(m) = \frac{1}{N} \sum_{n=1}^{N} D(n, m), \tag{2.42}$$

where $\bar{D}(m)$ is the average of measurement (m), and if

$$\sigma_{D(m)} = \left\{ \frac{1}{N-1} \sum_{n=1}^{N} [D(n, m) - \bar{D}(m)]^2 \right\}^{1/2}, \tag{2.43}$$

where $\sigma_{D(m)}$ is the standard deviation of measurement (m), then

$$SST = \sum_{m=1}^{M} \sum_{n=1}^{N} \left(\frac{D(n, m) - \bar{D}(m)}{\sigma_{D(m)}} \right)^2 \tag{2.44}$$

and

$$SSE = \sum_{m=1}^{M} \sum_{n=1}^{N} \left(\frac{E(n, m)}{\sigma_{D(m)}} \right)^2. \tag{2.45}$$

SST and SSE are then used in Equation (2.41) to calculate the multiple correlation coefficient. Equation (2.44) can be simplified by substitution of Equations (2.42) and (2.43):

$$SST = M(N - 1). \tag{2.46}$$

Figure 2-13 summarizes the validation procedure. This flow diagram is based on the general time simulation diagram (Figure 2-1), appropriately modified to accomplish the validation.

Specific Example

The Philadelphia Water Department has made available operating data recorded by their data logging computer. A more complex district—the Torresdale High Service-Fox Chase Booster District—is used here for the purpose of modeling and control studies. The district comprises twenty-seven square miles, with elevations varying from the zero city datum to 253 feet above the datum. Figure 2-14 shows a map of the pressure district along with an approximate scale included to give an indication of the distance separating pumping stations and tanks. The district is primarily residential, having a daily average consumption of about 30 million gallons. Figures 2-15 and 2-16 show the pump curves for the Torresdale and Fox Chase pumping stations, respectively. The Torresdale station contains five pumps, so many more combinations are possible than the five shown, which are the most frequently used. The Fox Chase station has two pump sizes.

Macroscopic Model Produced from a Dynamic Network Model Simulation

A network model is made to produce simulated time varying operating data as shown generally by Figure 2-1. The purpose for carrying out that simulation and the subsequent construction of a macroscopic model is to show by numerical example the accuracy with which the macroscopic model can

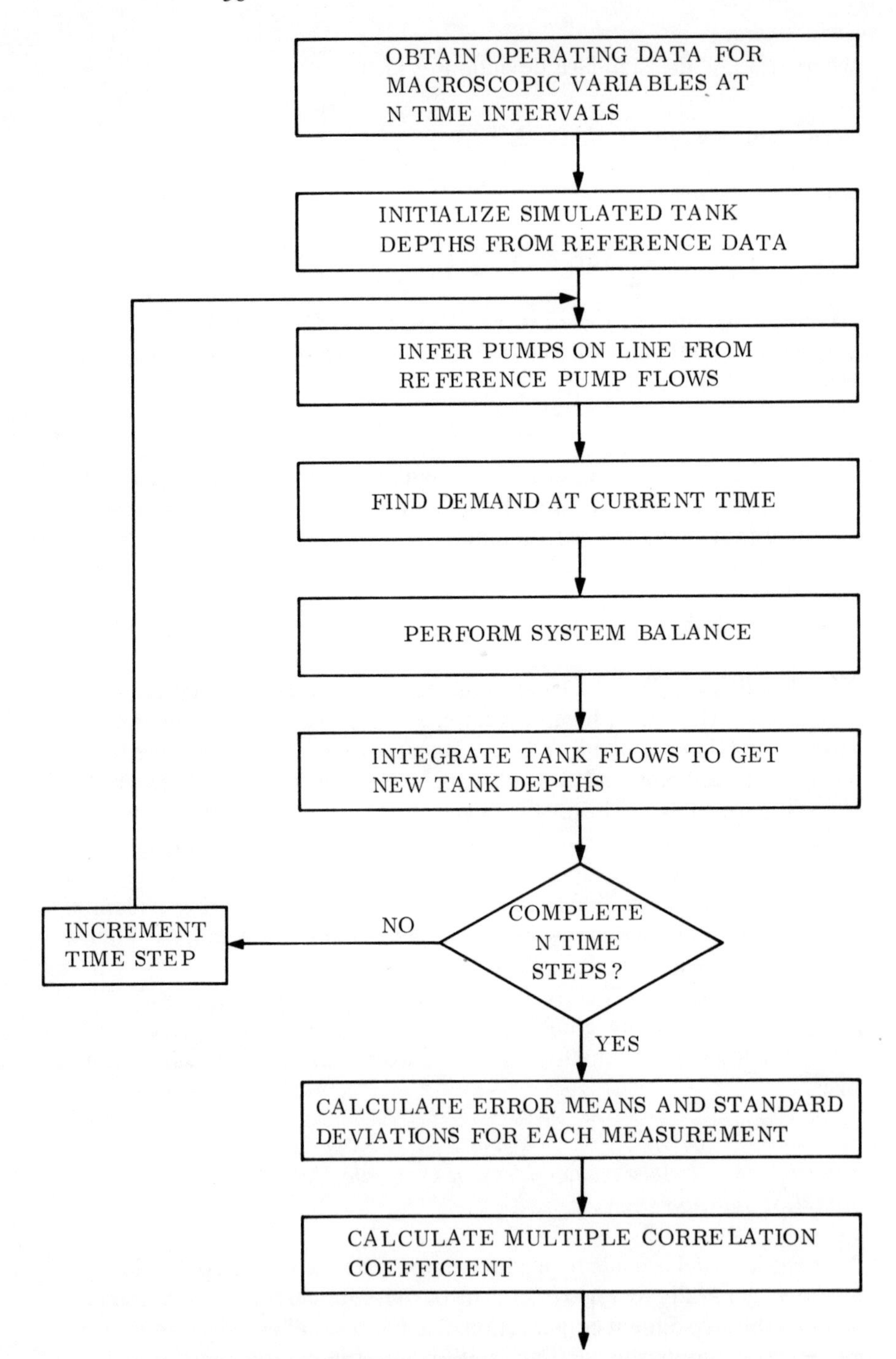

Figure 2-13. Model Validation by Time Simulation

P1 : TORRESDALE PUMPING STATION
P2 : FOX CHASE PUMPING STATION
T1 : SOMERTON TANK
T2 : FOX CHASE TANK
PP1: PRESSURE, PINE AND STANWOOD
PP2: PRESSURE, BUSTLETON & BOWLER

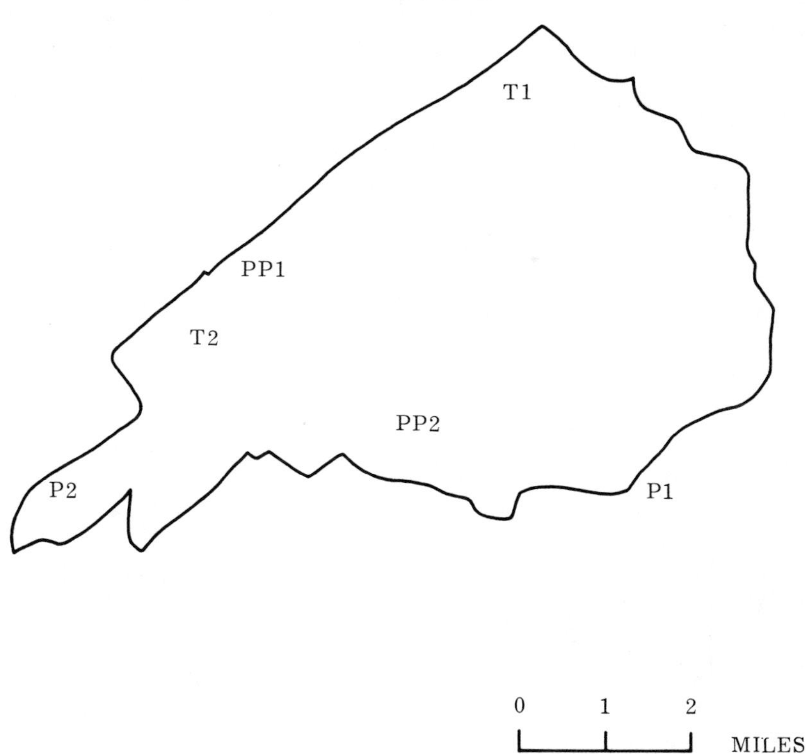

Figure 2-14. Torresdale High Service-Fox Chase Booster District

reproduce the simulated measurements. Figure 2-17 shows the total demand forcing function that is applied proportionally to the modes in a General Electric [18] network model. The total demand function contains up to the sixth harmonic of one day with no random noise. The full network model is presumably a representation of the Torresdale-Fox Chase district but is several years out of date, so the operating data simulated by the network are not representative of current operating data.

Table 2-2 shows data generated by simulated supervisory control carried out by a dynamic network model simulation. The pump changes were simulated somewhat arbitrarily as the result of watching tank depths and flows printed out on a time sharing computer. The interval of integration is

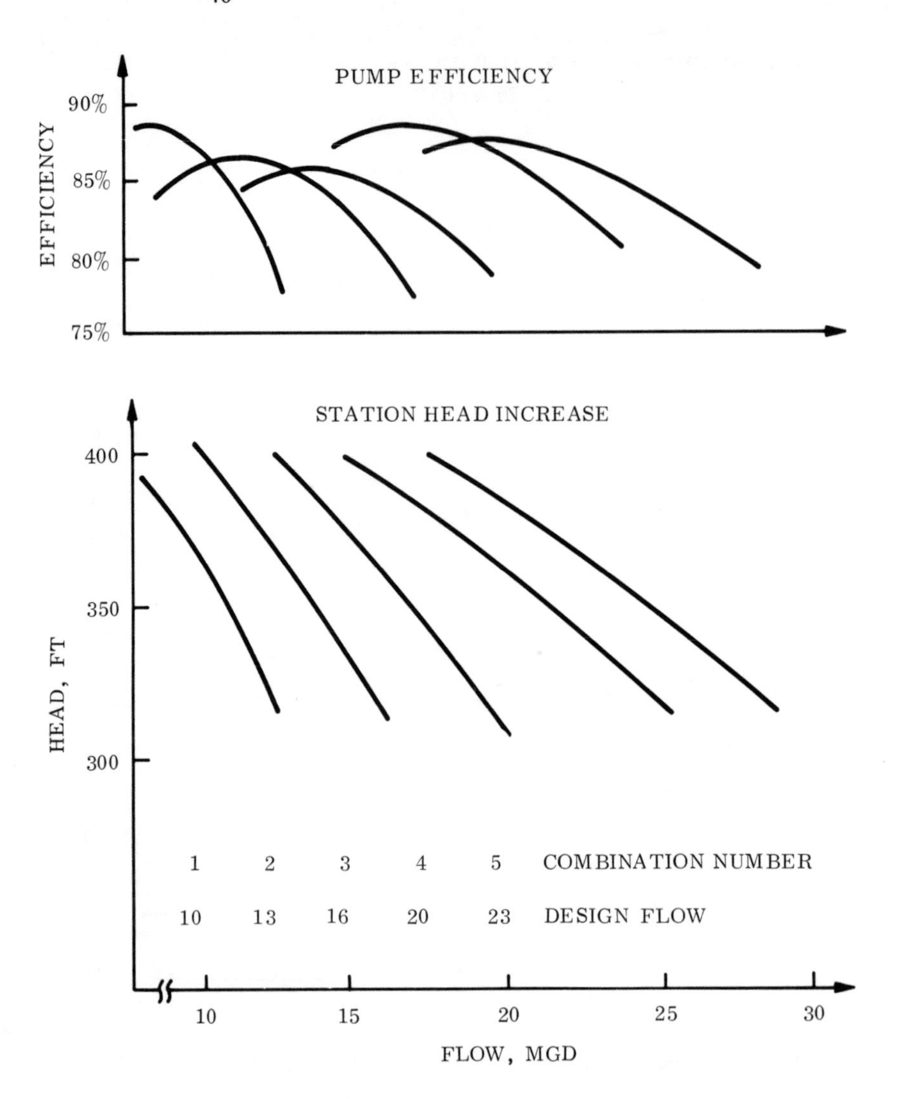

Figure 2-15. Torresdale Pump Curves

the same as the data logger recording interval—one-half hour. This interval is adequate since the period of the highest harmonic in the forcing function is four hours and since a typical time constant is several days.

Pump flows are shown in Figure 2-18, where the discontinuities are the result of simulated pump changes at 06:30, 10:30, and 20:00 hours. The effect of the two major pump changes at 06:30 and 20:00 are noticeable in the tank head plots in Figure 2-19. The Somerton tank (tank 1) is eighty-five feet tall, but to keep the system balanced, the operating range is the upper

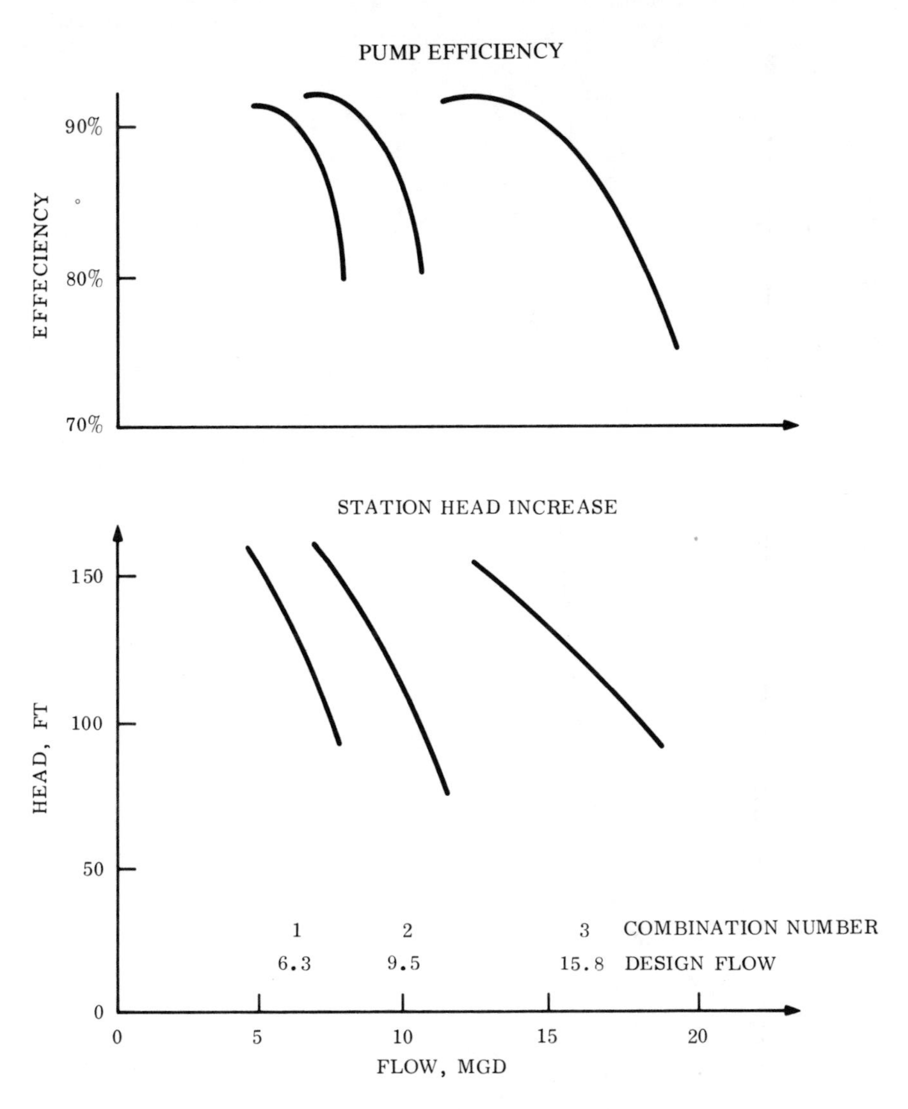

Figure 2-16. Fox Chase Pump Curves

twenty-five feet corresponding to the twenty-five foot depth of the Fox Chase tank.

A macroscopic model was constructed according to the procedure outlined in the section on "Summary of Macroscopic Model Construction" and the model was validated according to the section on "Model Validation," which appeared earlier in this chapter. Table 2-2 is the reference array $D(n, m)$. In the validation, the macroscopic model produces simulated measurements $\hat{D}(n, m)$. Rather than show these, the table of errors

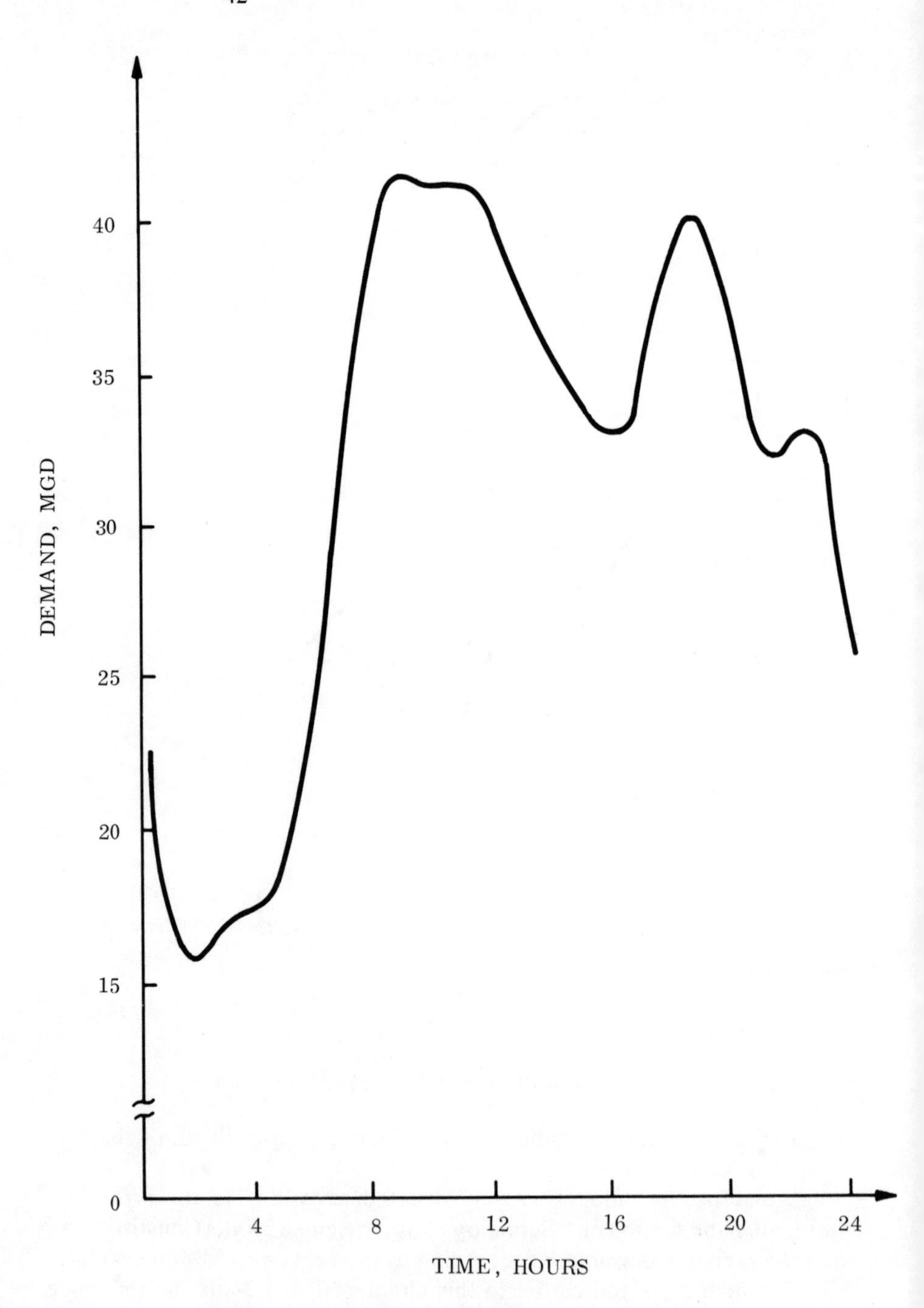

Figure 2-17. Total Demand Forcing Function

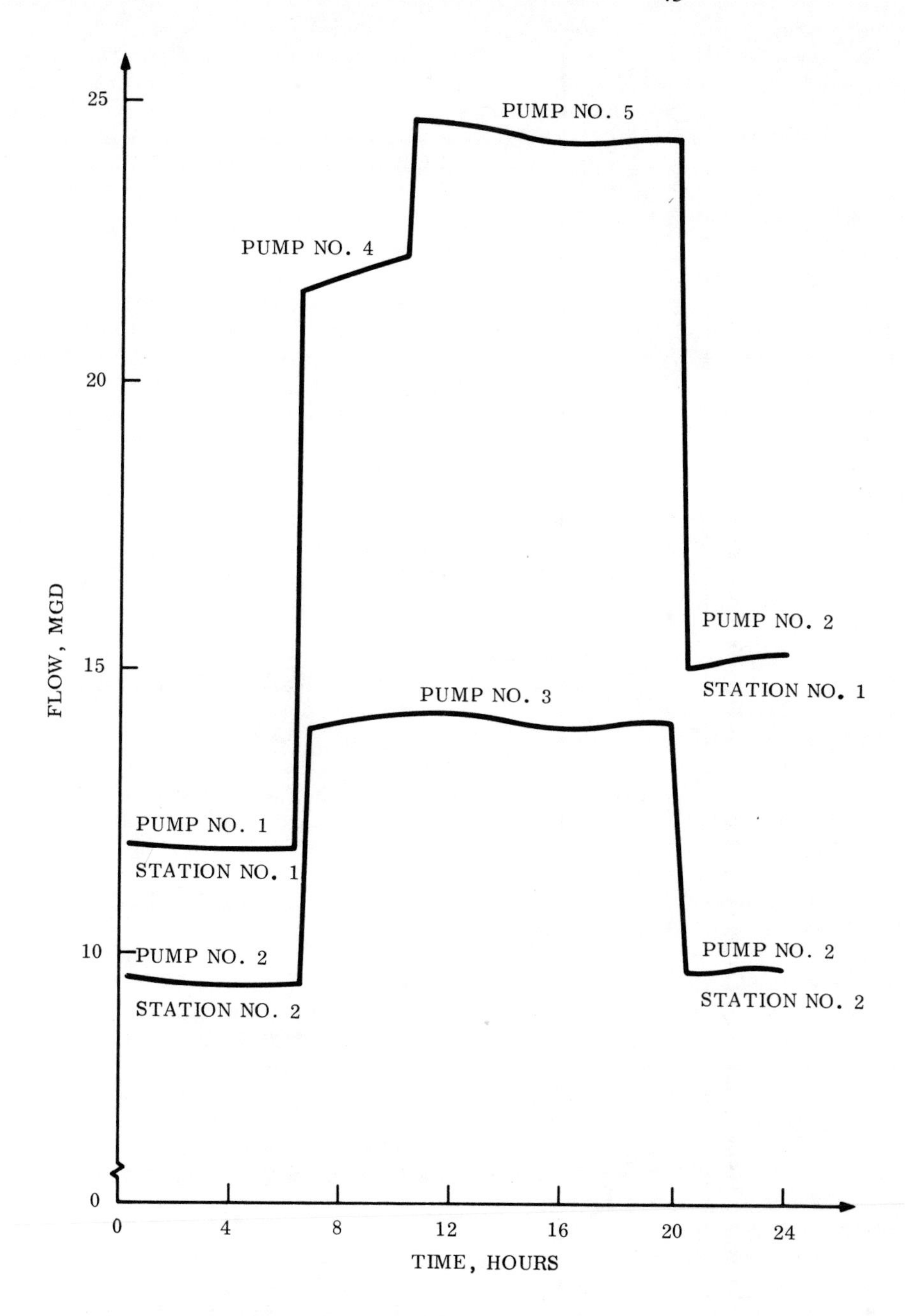

Figure 2-18. Simulated Pump Flow

Table 2-2
Dynamic Network Model Simulated Data: Forty-Eight Half-Hourly Time Points

Pump 1 discharge pressure (psi)[a]	*Pump 2 discharge pressure (psi)*[a]	*Pump 1 flow (mgd)*[b]	*Pump 2 flow (mgd)*[b]	*Tank 1 depth (ft)*	*Tank 2 depth (ft)*	*Tank 1 flow (mgd)*[b]	*Tank 2 flow (mgd)*[b]	*Internal pressure at pressure point 1 (psi)*[a]	*Internal pressure at pressure point 2 (psi)*[a]
145.9	84.7	11.9	9.5	76.9	11.5	−2.4	2.2	75.8	46.2
146.0	86.0	11.9	9.4	76.5	12.1	0.2	2.8	76.0	47.0
146.3	86.9	11.8	9.4	76.5	12.9	1.7	3.0	76.2	47.6
146.5	87.3	11.8	9.3	76.8	13.8	2.1	3.0	76.5	47.9
146.5	87.3	11.8	9.3	77.1	14.7	1.8	2.8	76.5	48.0
146.6	87.3	11.8	9.3	77.4	15.5	1.4	2.6	76.5	48.1
146.6	87.3	11.8	9.3	77.6	16.3	1.3	2.4	76.6	48.2
146.7	87.5	11.8	9.3	77.8	17.0	1.2	2.3	76.7	48.4
146.8	87.5	11.8	9.3	78.0	17.7	1.0	2.1	76.7	48.5
146.7	87.2	11.8	9.3	78.2	18.3	0.1	1.7	76.7	48.4
146.5	86.5	11.8	9.3	78.2	18.8	−1.9	0.8	76.5	48.1
146.0	85.6	11.9	9.4	77.9	19.0	−4.5	−0.6	76.0	47.6
152.3	94.6	21.5	13.7	77.2	18.9	2.0	2.1	76.8	48.2
151.8	93.2	21.7	13.9	77.5	19.5	−0.8	0.8	76.3	47.3
151.5	92.3	21.8	13.9	77.4	19.7	−2.9	−0.4	76.0	46.8
151.1	91.7	21.9	14.0	76.9	19.6	−4.0	−1.0	75.6	46.4
150.8	91.3	21.9	14.0	76.3	19.3	−4.4	−1.2	75.2	46.0
150.6	91.3	22.0	14.1	75.6	19.0	−4.3	−1.2	75.0	45.9
150.4	91.3	22.0	14.1	75.0	18.6	−4.0	−1.2	74.8	45.8
150.2	91.2	22.1	14.1	74.3	18.3	−3.9	−1.1	74.6	45.6

150.0	91.1	22.2	14.1	73.7	17.9	−3.8	−1.2	74.3	45.4
149.8	91.1	22.2	14.1	73.1	17.6	−3.6	−1.1	74.1	45.3
149.7	91.2	22.2	14.1	72.5	17.3	−3.2	−1.0	74.0	45.3
149.6	91.4	22.3	14.1	72.0	17.0	−2.5	−0.7	73.9	45.4
149.6	91.7	22.3	14.1	71.7	16.8	−1.6	−0.3	73.8	45.5
149.6	92.0	22.3	14.1	71.4	16.7	−0.9	0.0	73.8	45.6
149.7	92.3	22.2	14.1	71.3	16.7	−0.2	0.3	73.9	45.7
149.8	92.5	22.2	14.1	71.2	16.8	0.3	0.6	74.0	45.9
149.9	92.8	22.2	14.0	71.3	17.0	0.8	0.8	74.1	46.1
150.1	93.1	22.1	14.0	71.4	17.2	1.3	1.0	74.2	46.3
150.2	93.4	22.1	13.9	71.6	17.5	1.7	1.1	74.4	46.6
150.3	93.5	22.1	13.9	71.9	17.8	1.7	1.1	74.5	46.7
150.2	93.1	22.1	13.9	72.1	18.1	1.2	0.9	74.4	46.5
150.1	92.8	22.1	14.0	72.3	18.4	0.2	0.4	74.2	46.3
149.9	92.3	22.2	14.0	72.4	18.5	−1.1	−0.3	74.1	46.1
149.6	91.7	22.3	14.0	72.2	18.4	−2.3	−0.9	73.8	45.7
149.4	91.3	22.3	14.1	71.8	18.2	−2.8	−1.1	73.6	45.4
149.4	91.5	22.3	14.1	71.4	17.8	−2.5	−1.0	73.6	45.5
149.4	92.0	22.3	14.1	71.0	17.5	−1.2	−0.4	73.6	45.7
143.7	81.4	15.2	9.7	70.8	17.4	−7.2	−3.6	72.3	43.6
143.7	82.0	15.2	9.7	69.7	16.3	−5.6	−3.0	72.3	44.0
143.5	82.1	15.2	9.7	68.8	15.5	−4.8	−2.7	72.1	44.1
143.2	81.8	15.2	9.7	68.1	14.7	−4.8	−2.7	71.8	43.7
142.8	81.3	15.3	9.7	67.3	13.9	−5.1	−2.8	71.3	43.1
142.4	80.9	15.4	9.8	66.5	13.1	−5.3	−2.8	70.9	42.7
142.3	81.1	15.4	9.8	65.7	12.2	−4.5	−2.6	70.8	42.8
142.3	81.7	15.4	9.8	65.0	11.5	−2.7	−1.9	70.8	43.2
142.5	82.7	15.4	9.7	64.5	10.9	−0.0	−0.8	70.9	43.8

[a]The abbreviation psi stands for pounds per square inch.
[b]The abbreviation mgd stands for million gallons per day.

Table 2-3
Validation Errors Relative to Network Simulation Data: Forty-Eight Half-Hourly Points

Pump 1 discharge pressure (psi)[a]	*Pump 2 discharge pressure (psi)*[a]	*Pump 1 flow (mgd)*[b]	*Pump 2 flow (mgd)*[b]	*Tank 1 depth (ft)*	*Tank 2 depth (ft)*	*Tank 1 flow (mgd)*[b]	*Tank 2 flow (mgd)*[b]	*Internal pressure at pressure point 1 (psi)*[a]	*Internal pressure at pressure point 2 (psi)*[a]
0.1	−0.2	−0.0	0.1	0.	0.	0.6	−0.6	0.0	0.2
0.2	−0.4	−0.0	0.1	0.1	−0.1	0.2	−0.3	0.0	−0.0
0.0	−0.7	0.0	0.0	0.2	−0.2	−0.0	0.1	0.0	−0.2
−0.0	−0.6	0.0	0.1	0.1	−0.2	−0.1	0.1	−0.1	−0.2
0.1	−0.4	0.0	0.1	0.2	−0.2	−0.1	0.1	−0.0	−0.1
0.0	−0.2	−0.0	0.0	0.1	−0.1	−0.1	0.0	0.1	−0.1
0.1	−0.0	−0.0	0.0	0.1	−0.1	−0.0	0.1	0.0	−0.0
0.1	−0.0	−0.0	−0.0	0.1	−0.1	−0.0	0.1	0.0	−0.0
0.0	0.2	−0.0	−0.0	0.1	−0.1	−0.1	0.1	0.1	0.0
0.1	0.4	−0.0	−0.0	0.0	−0.0	−0.1	0.0	0.0	0.1
0.0	0.5	0.0	−0.0	0.0	−0.0	0.0	−0.0	0.0	0.1
−0.0	0.3	−0.0	−0.0	0.1	−0.0	−0.0	−0.0	−0.0	−0.1
0.0	0.2	0.0	−0.0	0.0	−0.1	0.0	−0.1	0.1	0.3
0.1	0.4	−0.1	−0.1	0.1	−0.1	−0.0	−0.1	0.2	0.5
−0.0	0.1	−0.0	0.0	0.0	−0.1	−0.0	0.0	0.1	0.1
−0.0	−0.1	−0.0	0.0	0.1	−0.1	−0.0	0.1	−0.0	−0.0
−0.0	0.0	0.0	0.1	0.0	−0.1	0.0	0.1	0.0	0.1
−0.1	−0.1	0.0	−0.0	0.0	−0.0	0.0	0.1	−0.1	0.0
−0.1	−0.1	0.1	−0.0	−0.0	−0.0	−0.1	0.1	−0.1	−0.0
−0.1	−0.1	0.0	−0.0	0.0	−0.1	−0.0	0.0	−0.0	−0.1

0.1	−0.2	−0.0	0.1	0.0	0.0	−0.1	0.1	0.0	−0.1
0.0	−0.3	−0.0	0.1	−0.0	0.0	−0.1	0.1	−0.1	−0.2
0.0	−0.4	−0.0	0.1	0.0	0.1	−0.0	0.1	−0.1	−0.3
0.1	−0.2	−0.0	0.0	−0.1	0.1	−0.0	0.1	−0.1	−0.1
0.1	−0.1	0.0	−0.0	−0.0	0.1	−0.1	0.0	0.1	0.0
0.0	0.1	−0.0	−0.1	−0.0	0.1	0.1	−0.0	0.0	0.1
−0.0	0.1	0.0	0.0	−0.1	0.1	0.1	−0.1	0.0	0.2
−0.0	0.1	−0.0	−0.0	−0.0	0.1	0.1	−0.1	0.0	0.1
−0.0	−0.0	0.0	−0.1	0.0	0.0	0.1	−0.1	−0.1	0.1
−0.1	0.0	0.1	−0.0	0.0	0.0	0.0	−0.0	−0.1	0.1
−0.1	0.0	0.0	−0.1	−0.0	0.0	0.1	−0.1	−0.2	0.0
−0.1	−0.1	0.0	0.0	−0.0	0.0	0.1	−0.1	−0.2	−0.1
−0.0	0.2	0.0	−0.1	−0.0	−0.1	0.0	−0.1	0.1	0.2
0.1	0.3	−0.1	−0.1	0.0	0.0	−0.0	−0.1	0.2	0.2
0.1	0.1	−0.0	0.0	−0.0	−0.1	−0.1	0.0	0.1	0.0
0.2	−0.1	−0.1	−0.0	0.0	−0.1	−0.1	0.1	0.1	−0.1
0.1	−0.2	−0.0	0.0	−0.0	0.0	−0.1	0.1	0.1	−0.1
0.1	−0.3	−0.0	0.0	−0.0	0.0	−0.1	0.1	−0.1	−0.3
0.1	−0.1	−0.1	−0.0	−0.0	0.0	−0.0	−0.0	−0.1	−0.0
0.1	0.1	−0.0	-0.0	0.0	0.0	0.2	−0.2	0.1	0.3
0.0	0.0	0.0	−0.0	0.0	−0.0	0.0	−0.1	−0.1	0.0
−0.0	0.1	0.0	−0.0	0.0	−0.0	0.1	−0.1	−0.2	−0.1
−0.1	0.1	−0.0	−0.0	−0.0	−0.1	0.1	−0.1	−0.1	0.1
−0.1	0.1	0.0	0.0	0.1	−0.1	0.0	−0.0	0.1	0.1
−0.1	0.1	0.0	−0.0	0.1	−0.1	0.0	−0.0	0.1	0.3
−0.1	−0.0	0.0	−0.0	0.1	−0.1	0.0	0.0	0.0	0.1
−0.1	0.1	0.0	−0.1	0.1	−0.1	−0.1	0.1	0.0	−0.0
−0.0	0.2	0.0	−0.0	0.1	−0.0	−0.3	0.2	0.1	−0.0

[a]The abbreviation psi stands for pounds per square inch.
[b]The abbreviation mgd stands for million gallons per day.

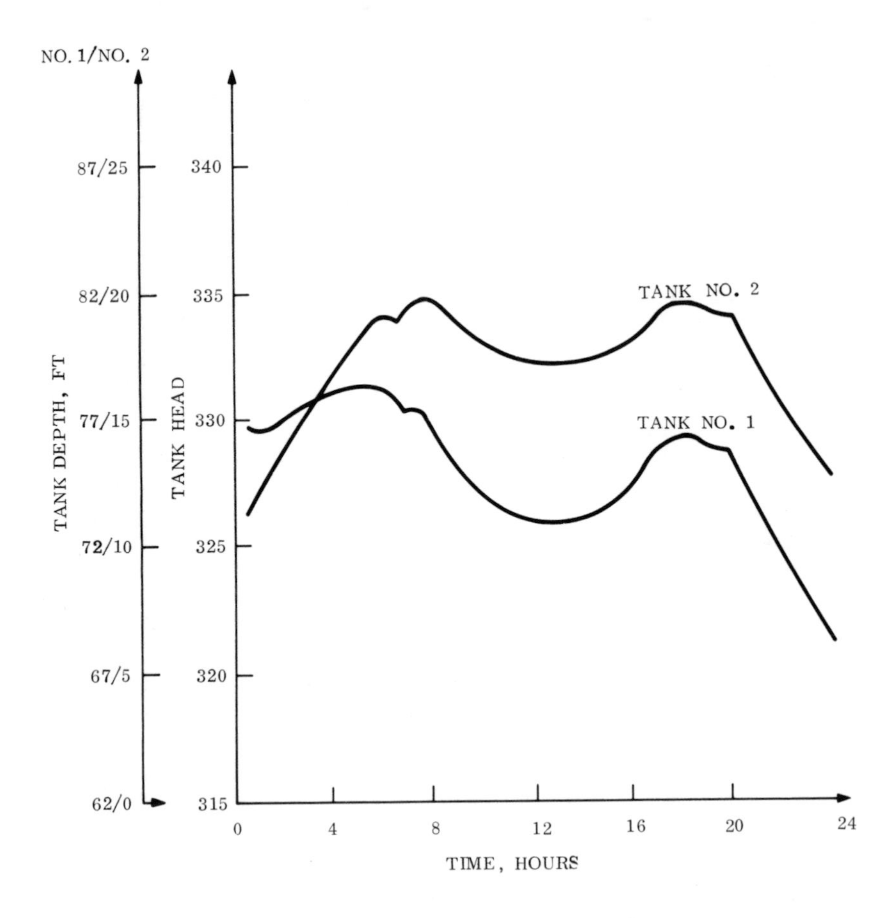

Figure 2-19. Simulated Tank Depth

$E(n, m)$ [Equation (2.38)] is shown in Table 2-3. The summary of error statistics is shown in Table 2-4.

It can be concluded that a macroscopic model time simulation can be essentially equivalent to a dynamic network model simulation under proportional loading. However, they are not completely equivalent. In studying McPherson's generalized head drop expression, Davis [15] presented a simple proportional loading example with equalizing storage. The example showed that McPherson's expression did not completely describe the network head drop. This example applies equally well to show that Equation (2.6) is not completely adequate. The problem is caused by the fact that, while all node consumptions vary proportionally, pipe flows in the vicinity of the tanks do not. In fact, Table 2-2 shows that tank flows actually reverse from time to time. The reason for the success of the macroscopic model is that across the miles separating the pumping stations and tanks,

Table 2-4
Validation Error Statistics Relative to Dynamic Network Simulation[a]

Measurement	*Average*	*Standard deviation*
Pump 1 discharge pressure	0.01 psi[b]	0.08
Pump 2 discharge pressure	−0.03 psi[b]	0.24
Pump 1 flow	−0.00 mgd[c]	0.03
Pump 2 flow	−0.00 mgd[c]	0.05
Tank 1 depth	0.04 ft	0.06
Tank 2 depth	−0.03 ft	0.08
Pressure 1	−0.00 psi[b]	0.10
Pressure 2	0.03 psi[b]	0.16

[a]Multiple correlation coefficient—0.9981.
[b]The abbreviation psi stands for pounds per square inch.
[c]The abbreviation mgd stands for million gallons per day.

the majority of flows are substantially proportional. In the vicinity of tanks where flows are not proportional, pipe diameters are large, so head loss is small.

Through use of the very fast General Electric network balance program, [24], about five minutes of central processor time were required to carry out the forty-eight step dynamic network simulation. The macroscopic model carried out the same simulation in about two seconds. The time reduction is largely the result of the elimination of calculation of hundreds of unnecessary flows and pressures at each time interval.

Macroscopic Model Produced from Actual Operating Data

The previous section demonstrated that the macroscopic model can be an efficient replacement for a network model for the purpose of control. In practice, however, loading is not truly proportional. The purpose of this section is to show the ability of the macroscopic model to simulate actual operating data.

Table 2-5 shows actual operating data from Monday, 10 January 1972. To overcome tank flow measurement inaccuracies, the tank flows shown were calculated from the change in tank depth, and then total demand was calculated from flow continuity. A macroscopic model was constructed from this data, and then the model was validated. The table of errors between macroscopic model simulated data and actual data appears in Table 2-6. The error summary appears in Table 2-7, and a comparison of simulated to actual tank depths is shown in Figure 2-20.

Table 2-5
Operating Data: 10 January 1972

Pump 1 discharge pressure (psi)[a]	*Pump 2 discharge pressure (psi)*[a]	*Pump 1 flow (mgd)*[b]	*Pump 2 flow (mgd)*[b]	*Tank 1 depth (ft)*	*Tank 2 depth (ft)*	*Tank 1 flow (mgd)*[b]	*Tank 2 flow (mgd)*[b]	*Internal pressure at pressure point 1 (psi)*[a]	*Internal pressure at pressure point 2 (psi)*[a]
144.0	83.5	11.9	10.0	76.9	11.5	−1.3	2.7	73.3	48.0
144.0	83.6	11.8	10.0	76.7	12.3	3.8	2.4	73.1	48.3
144.0	84.3	11.8	10.1	77.3	13.0	−1.3	2.4	73.7	48.9
144.0	84.8	11.8	10.0	77.1	13.7	3.8	2.4	73.4	49.4
144.0	85.4	11.9	9.9	77.7	14.4	2.5	2.0	74.1	49.5
145.0	85.5	11.8	9.9	78.1	15.0	4.5	2.7	73.7	49.8
145.0	85.6	11.9	9.9	78.8	15.8	2.5	1.7	74.8	50.1
144.0	85.5	11.8	9.8	79.2	16.3	2.5	2.4	74.2	49.7
145.0	85.8	11.8	9.8	79.6	17.0	1.9	1.7	74.7	50.3
145.0	85.7	11.8	9.8	79.9	17.5	−0.6	1.7	74.5	50.7
145.0	85.2	11.8	9.8	79.8	18.0	−3.2	0.7	74.6	50.8
145.0	84.8	11.9	9.7	79.3	18.2	0.	0.	74.1	50.3
150.0	84.2	22.2	9.9	79.3	18.2	2.5	−1.0	75.4	50.0
149.0	83.3	22.3	9.8	79.7	17.9	−3.2	−2.0	75.3	49.0
148.0	81.8	22.3	9.8	79.2	17.3	−5.7	−3.1	73.2	47.7
148.0	81.6	22.4	9.9	78.3	16.4	−5.1	−2.4	73.6	47.2
148.0	81.5	22.5	9.8	77.5	15.7	−4.5	−3.1	74.0	47.2
149.0	80.4	22.3	9.9	76.8	14.8	−5.1	−3.1	72.7	47.0
148.0	80.3	22.1	9.9	76.0	13.9	−7.0	−3.4	72.4	47.0
146.0	78.6	22.3	9.9	74.9	12.9	−7.0	−3.4	71.5	44.7

145.0	78.8	22.7	9.9	73.8	11.9	−6.4	−2.7	70.8	44.6
146.0	79.2	22.5	10.0	72.8	11.1	−3.8	−3.4	70.7	44.8
144.0	78.9	22.9	10.0	72.2	10.1	−3.2	−2.4	70.5	43.4
144.0	78.4	25.5	10.0	71.7	9.4	−2.5	−0.3	70.1	43.8
147.0	78.7	25.4	9.9	71.3	9.3	−1.9	−3.7	70.7	44.3
147.0	79.1	25.0	9.9	71.0	8.2	−0.6	−1.7	68.5	45.2
146.0	79.5	25.2	9.9	70.9	7.7	0.	−1.0	69.5	45.1
147.0	78.8	25.3	9.9	70.9	7.4	−1.3	1.4	70.1	45.7
146.0	88.9	25.4	14.6	70.7	7.8	5.1	3.1	70.7	45.3
146.0	89.2	25.5	14.6	71.5	8.7	3.2	2.7	70.4	46.6
148.0	88.8	25.5	14.5	72.0	9.5	2.5	2.7	72.1	47.2
148.0	89.1	25.5	14.4	72.4	10.3	5.1	2.4	73.0	46.3
149.0	90.0	25.4	14.3	73.2	11.0	2.5	2.4	70.9	47.1
147.0	89.7	25.7	14.4	73.6	11.7	3.2	2.4	71.7	47.1
149.0	89.3	25.7	14.4	74.1	12.4	1.9	1.4	72.3	47.5
147.0	89.1	25.8	14.4	74.4	12.8	−0.6	1.4	72.0	47.9
148.0	89.7	25.6	14.3	74.3	13.2	0.	0.7	72.0	47.7
147.0	89.0	25.7	14.3	74.3	13.4	−1.3	1.4	71.8	45.5
147.0	89.4	25.8	14.3	74.1	13.8	−0.6	−0.7	71.8	47.4
148.0	80.8	25.7	9.7	74.0	13.6	6.4	−1.4	71.6	46.4
148.0	81.5	25.5	9.8	75.0	13.2	0.	−1.0	72.4	47.3
148.0	80.0	25.6	9.8	75.0	12.9	3.2	−1.0	71.7	45.9
148.0	81.7	25.6	9.8	75.5	12.6	5.7	−0.3	72.5	47.5
149.0	81.7	25.4	9.8	76.4	12.5	3.8	0.7	72.8	48.2
151.0	83.5	25.3	9.8	77.0	12.7	0.6	1.0	75.3	49.3
142.0	78.8	18.0	9.8	77.1	13.0	−8.9	−1.7	69.8	45.0
144.0	81.5	18.0	9.8	75.7	12.5	0.6	1.0	71.9	47.1
145.0	82.9	17.8	9.8	75.8	12.8	1.3	0.7	72.5	47.3

[a]The abbreviation psi stands for pounds per square inch.
[b]The abbreviation mgd stands for million gallons per day.

Table 2-6
Validation Errors Relative to Operating Data: Forty-Eight Half-Hourly Steps

Pump 1 discharge pressure (psi)[a]	*Pump 2 discharge pressure (psi)*[a]	*Pump 1 flow (mgd)*[b]	*Pump 2 flow (mgd)*[b]	*Tank 1 depth (ft)*	*Tank 2 depth (ft)*	*Tank 1 flow (mgd)*[b]	*Tank 2 flow (mgd)*[b]	*Internal pressure at pressure point 1 (psi)*[a]	*Internal pressure at pressure point 2 (psi)*[a]
−0.2	−0.7	0.3	−0.2	0.	0.	0.7	−0.7	−0.4	0.3
−0.1	0.0	0.3	−0.3	0.1	−0.2	−0.5	0.6	0.0	0.5
−0.0	−0.1	0.3	−0.4	0.0	−0.0	0.7	−0.8	−0.5	−0.3
0.1	−0.7	0.3	−0.3	0.1	−0.3	−0.3	0.3	0.0	−0.3
0.3	−1.2	0.2	−0.3	0.1	−0.2	−0.2	0.2	−0.4	−0.3
−0.5	−0.7	0.3	−0.3	0.0	−0.1	−0.1	0.1	0.3	−0.1
−0.3	−0.9	0.1	−0.3	0.0	−0.1	−0.4	0.2	−0.5	−0.3
0.9	−0.5	0.2	−0.2	−0.0	−0.0	0.3	−0.4	0.3	0.4
0.0	−0.7	0.2	−0.3	0.0	−0.1	−0.0	−0.1	−0.0	−0.1
0.1	−0.7	0.2	−0.3	0.0	−0.1	0.7	−0.7	0.2	−0.5
−0.1	−0.6	0.2	−0.3	0.1	−0.3	0.6	−0.7	−0.1	−1.0
−0.1	0.1	0.1	−0.2	0.2	−0.5	−0.6	0.5	0.4	−0.4
−1.2	−0.7	0.3	−0.4	0.1	−0.4	−0.3	0.2	−0.4	−0.2
−0.4	−1.2	0.2	−0.2	0.1	−0.3	0.3	−0.3	−0.7	−0.1
0.2	−0.8	0.3	−0.2	0.1	−0.4	0.1	0.0	0.7	0.3
−0.1	−0.7	0.3	−0.3	0.1	−0.4	0.4	−0.3	−0.0	0.5
−0.4	−0.9	0.3	−0.1	0.2	−0.5	−0.2	0.4	−0.9	0.1
−1.7	−0.2	0.6	−0.2	0.2	−0.4	0.1	0.4	−0.0	−0.2
−1.1	−0.9	0.9	−0.1	0.2	−0.3	0.5	0.3	−0.4	−1.1
0.5	0.4	0.8	−0.1	0.3	−0.2	0.4	0.3	−0.1	0.7

1.1	0.1	0.5	−0.1	0.3	−0.1	0.5	−0.1	0.1	0.4
−0.1	−0.1	0.7	−0.2	0.4	−0.2	−0.5	1.1	−0.1	0.0
1.7	0.2	0.4	−0.1	0.3	0.1	−0.2	0.5	−0.2	1.3
3.1	0.5	0.4	−0.1	0.3	0.3	1.6	−1.3	0.3	1.0
−0.2	−0.5	0.6	0.0	0.5	−0.1	−1.0	1.6	−0.6	−0.1
−0.1	−0.3	1.0	0.0	0.4	0.4	0.5	0.6	1.7	−0.6
1.0	−0.6	0.8	0.0	0.4	0.5	0.6	0.2	0.8	−0.3
0.1	0.2	0.6	0.0	0.5	0.6	2.5	−1.9	0.3	−0.7
1.3	0.7	0.5	0.2	0.9	0.1	0.0	0.5	0.7	1.6
1.5	0.3	0.3	0.1	0.9	0.2	0.2	0.2	1.2	0.4
−0.3	0.8	0.3	0.2	1.0	0.3	0.6	−0.1	−0.2	0.0
−0.0	1.2	0.2	0.2	1.1	0.3	−0.3	0.7	−0.6	1.5
−0.9	0.1	0.2	0.3	1.0	0.5	0.6	−0.0	1.7	0.8
1.3	0.7	−0.1	0.1	1.1	0.5	0.1	−0.1	1.2	1.1
−0.6	0.8	−0.1	0.1	1.1	0.5	−0.3	0.3	0.7	0.6
1.3	0.5	−0.2	0.1	1.1	0.5	0.3	−0.5	0.9	−0.1
0.3	0.1	−0.0	0.2	1.1	0.4	−0.0	0.2	0.9	0.1
1.2	0.7	−0.1	0.2	1.1	0.5	0.8	−0.7	1.0	2.2
1.1	0.0	−0.2	0.2	1.2	0.3	−0.9	0.9	0.8	−0.0
0.7	1.4	−0.2	−0.0	1.1	0.5	−1.3	1.1	1.3	1.4
0.6	−0.5	−0.0	−0.1	0.9	0.8	0.5	−0.6	0.5	−0.0
0.8	1.5	−0.2	−0.1	1.0	0.7	−0.6	0.3	1.5	1.9
1.2	0.3	−0.3	−0.1	0.9	0.8	−0.9	0.5	1.1	1.0
0.4	0.2	−0.1	−0.1	0.8	0.9	0.3	−0.5	1.2	0.6
−1.5	−2.1	−0.1	−0.1	0.8	0.8	1.4	−1.4	−1.2	−0.6
3.6	1.5	0.1	−0.1	1.0	0.4	0.5	−0.5	3.0	2.1
1.7	0.9	0.0	−0.1	1.1	0.2	0.3	−0.2	1.2	1.1
0.8	−0.3	0.2	−0.1	1.1	0.2	−0.0	0.2	0.7	1.1

[a]The abbreviation psi stands for pounds per square inch.

[b]The abbreviation mgd stands for million gallons per day.

Table 2-7
Validation Error Statistics Relative to Operating Data[a]

Measurement	*Average*	*Standard deviation*
Pump 1 discharge pressure	0.35 psi[b]	1.02
Pump 2 discharge pressure	−0.08 psi[b]	0.78
Pump 1 flow	0.24 mgd[c]	0.30
Pump 2 flow	0.09 mgd[c]	0.17
Tank 1 depth	0.53 ft	0.67
Tank 2 depth	0.12 ft	0.40
Pressure 1	0.36 psi[b]	0.81
Pressure 2	0.32 psi[b]	0.79

[a]Multiple correlation coefficient—0.9467.
[b]The abbreviation psi stands for pounds per square inch.
[c]The abbreviation mgd stands for million gallons per day.

The error summary shows that average errors are small but that some error standard deviations are not. The model can simulate the overall system dynamics well, but it is not capable of reproducing the variations due to the random shifts in demand. Figure 2-21 shows the total demand forcing function. It is evident that actual demand contains harmonics higher than the sixth used in the dynamic network model simulation in Figure 2-17. However, it will be shown in Chapter 4 that only the first six harmonics are significant and that higher frequencies result from unaccountable random variations.

Two-Pump-Station/Two-Tank System Dynamics

Now that it has been established that the macroscopic model accurately reproduces operating data from the Torresdale-Fox Chase system, the model can be used to gain some insight into the system dynamics. The two-tank distribution system is particularly convenient to analyze because the tank depths (or heads) can be plotted against one another as state variables in a phase plane.

From Equation (2.13), tank depth rate is related to tank flow by the tank depth/volume characteristic:

$$d[Dt(j)] = Cdr(j)Qt\,dt\,;$$

or since, from Equation (2.7),

$$Ht(j) = He(j) + Dt(j),$$

Equation (2.13) can be equivalently written in terms of tank head:

$$d[Ht(j)] = Cdr(j)Qt\,dt\,. \qquad (2.47)$$

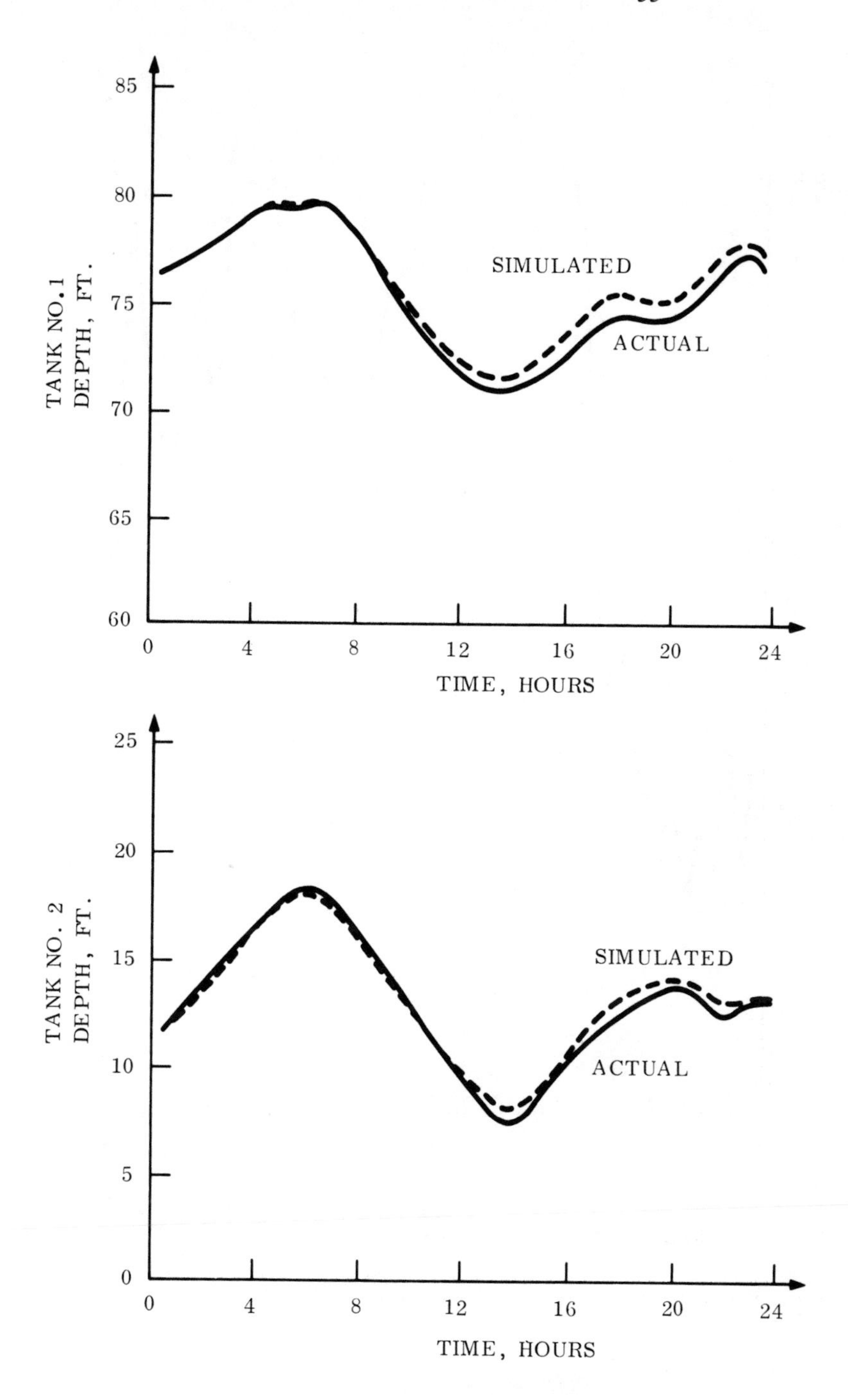

Figure 2-20. Comparison of Simulated to Actual Tank Depth

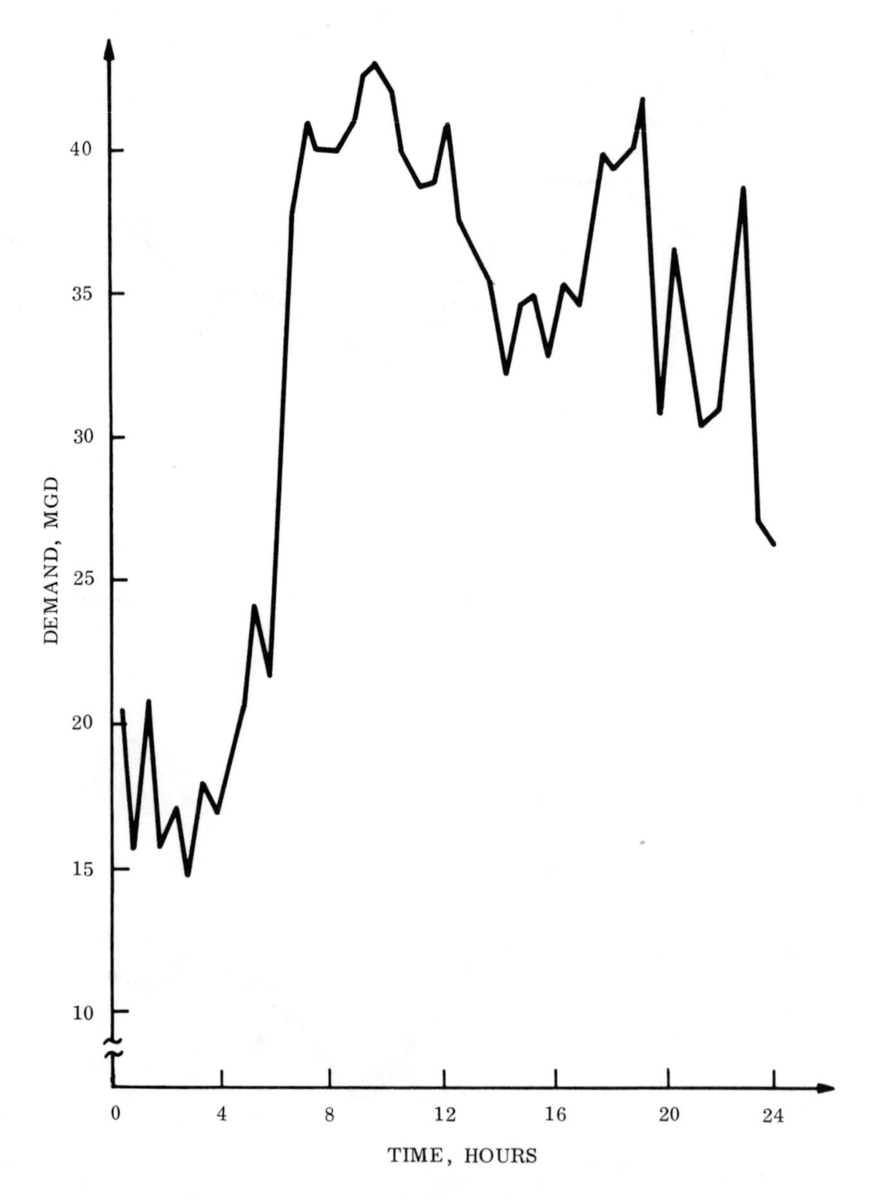

Figure 2-21. Total Demand for 10 January 1972

At any point in the phase plane, the slope is given by

$$\frac{d[Ht(2)]}{d[Ht(1)]} = \frac{Cdr(2)\,Qt(2)}{Cdr(1)\,Qt(1)}. \tag{2.48}$$

In the current example, the tank flow constants for tank 2 are calculated from those of tank 1. The phase plane [23] slope follows from the tank flow expression [Equation (2.8)]:

$$\frac{d[Ht(2)]}{d[Ht(1)]} = \frac{Ctdr(1)}{Ctdr(2)}$$

$$\left\{ \frac{-Ct(1,\,1) - [Ct(1,\,2) + 1]Qd - Ct(1,\,3)\,Ht(1)^{1/1.85}}{Ct(1,\,1) + Ct(1,\,2)Qd + Ct(1,\,3)Ht(1)^{1/1.85}} \right.$$

$$\left. \frac{-Ct(1,4)Ht(2)^{1/1.85} + [1 - Ct(1,5)]Qp(1) + [1 - Ct(1,6)]Qp(2)}{+Ct(1,4)Ht(2)^{1/1.85} + Ct(1,5)Qp(1) + Ct(1,6)Qp(2)} \right\}. \quad (2.49)$$

It is not possible to solve for $Ht(1)$ in terms of $Ht(2)$ and $d[Ht(2)]/d[Ht(1)]$, so the method of isoclines cannot be applied to the phase plane construction. However, owing to the speed of the macroscopic model, transient time responses can be made quite rapidly. Figure 2-22 shows a typical tank depth phase plane. For computational convenience, as each phase plane trajectory is calculated, the distance along the curve is calculated as a numerical line integral, causing the computer to print coordinates at equally spaced intervals. In order to give an indication of the speed of the response, marks along two of the trajectories indicate one-hour intervals.

The phase plane shown in Figure 2-22 shows that the equilibrium point is a stable node. This result is expected because of the miles of damping pipes connecting the pumps and tanks. Two asymptotic slopes are evident. First, when initial tank depths are far from the line of equal head, the primary effect is that the tank heads tend to equalize. Depending on the initial conditions, the water level may rise in one tank and fall in the other. In the latter stage of the transient, the tank heads remain substantially the same as the water levels in the tanks slowly rise or fall together to the equilibrium point.

We return now to the first part of the transient, where the demand flow and pump flows are relatively unimportant. If we set $Qd = Qp(1) = Qp(2) = 0$ in Equation (2.49), we obtain the slope, as follows:

$$\frac{d[Ht(2)]}{d[Ht(1)]_{\mathrm{I}}} \cong \frac{Cdr(2)}{Cdr(1)}$$

$$\left[\frac{-Ct(1,\,1) - Ct(1,\,3)Ht(1)^{1/1.85} - Ct(1,\,4)\,Ht(1,\,2)^{1/1.85}}{Ct(1,\,1) + Ct(1,\,3)\,Ht(1)^{1/1.85} + Ct(1,\,4)\,Ht(1,2)^{1/1.85}} \right], \quad (2.50)$$

or

$$\frac{d[Ht(2)]}{d[Ht(1)_{\mathrm{I}}} \cong -\frac{Cdr(2)}{Cdr(1)}. \quad (2.51)$$

In this specific case,

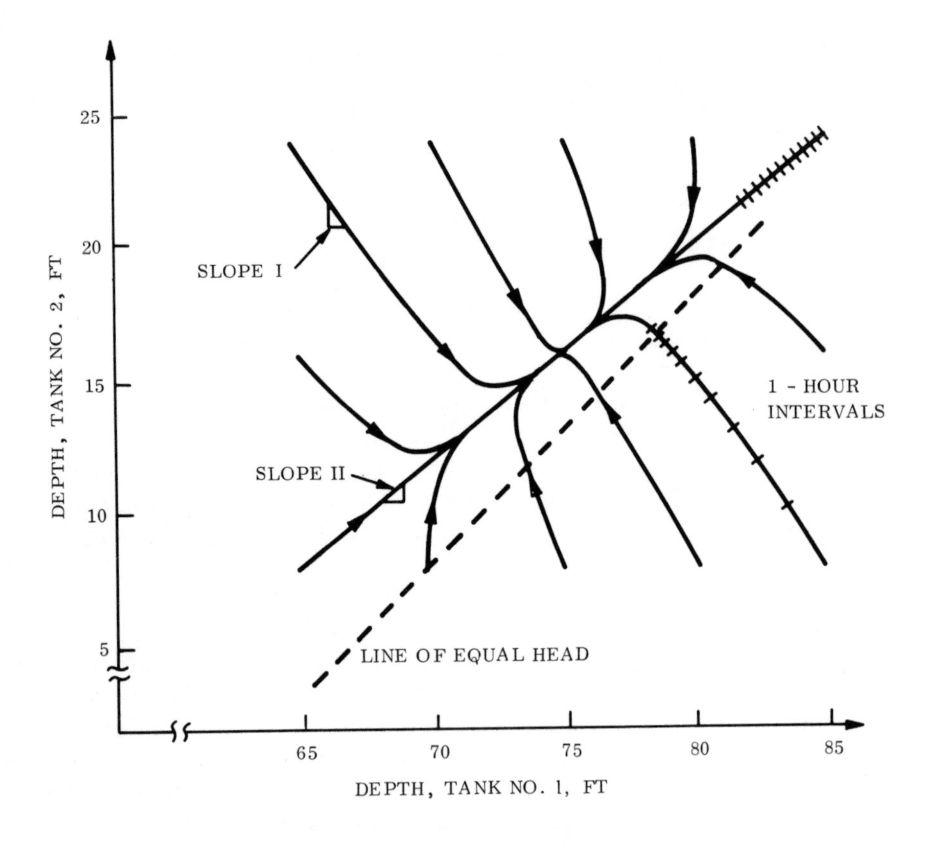

Figure 2-22. Tank Depth Phase Plane

$$\frac{d[Ht(2)]}{d[Ht(1)]} \cong -\frac{14.15}{7.54} = -1.88\,. \qquad (2.52)$$

To find the asymptotic slope in the second part of the transient, the flow into tank 1 [denominator of Equation (2.49)] is differentiated:

$$\begin{aligned} d[Qt(1)] = {} & Ct(1,2)d[Qd] + (1/1.85)Ct(1,3)Ht(1)^{(1/1.85-1)}\, d[Ht(1)] \\ & + (1/1.85)Ct(1,4)Ht(2)^{(1/1.85-1)}\, d[Ht(2)] + Ct(1,5)d[Qp(1)] \\ & + Ct(1,6)d[Qp(2)]\,. \qquad (2.53) \end{aligned}$$

Demand flow is constant, so $d[Qd] = 0$. As equilibrium is approached, pump flows become constant, so $d[Qp(1)] \cong d[Qp(2)] \cong 0$. Finally, tank flow becomes very small, so $d[Qt(1)] \cong 0$. Equation (2.53) is modified to reflect these conditions:

$$0 = (1/1.85)\, Ct(1, 3) Ht(1)^{(1/1.85-1)}\, d[Ht(1)] + (1/1.85)\, Ct(1, 4) Ht(2)^{(1/1.85-1)}\, d[Ht(2)]\,. \tag{2.54}$$

Since tank heads are very close, that is, since $TH(1) \cong TH(2)$,

$$\frac{d[Ht(2)]}{d[Ht(1)]_{\mathrm{II}}} \cong -\frac{Ct(1, 3)}{Ct(1, 4)}\,. \tag{2.55}$$

In this specific case,

$$\frac{d[Ht(2)]}{d[Ht(1)]_{\mathrm{II}}} \cong -\frac{-4.96}{6.41} = 0.78\,. \tag{2.56}$$

Asymptotic slope is a function of neither total demand or pump flow, so as demand changes or different pumps are put on line, the shape of the phase plane remains substantially the same. For a given set of pumps, the equilibrium point moves up the asymptotic slope II as demand decreases. For a given demand, the intercept of the asymptotic slope II rises up and down on the vertical axis as pump changes are made.

3 The Adaptive Model

Chapter 2 presented an algorithm for the rapid and accurate modeling of a water pressure district as it exists on one particular day. Distribution systems are subject to change for several reasons. Slow changes in hydraulic characteristics result from the aging of pipes through the buildup of deposits. More abrupt changes result from the opening of new pipes and the cleaning of mains. A full network model can be kept up to date only by programming changes, which can be quite awkward if pipes are added to the network when a loop formulation is used. Changes in pipe flow coefficients can be applied to a network model only after field measurements are made.

The macroscopic model can be kept current by the continual analysis of operating data. In addition to determining the results of changing hydraulics, the effects of changing loading patterns, caused by changing seasons, lead to changes in the head drop, tank flow, and pressure expressions. In looking ahead to control by dynamic programming, it is not necessary to have an up-to-date model at all times. The model, as it exists at one point in time, will be used to plan the control for several days into the future. For this reason, the requirement is to make available at the end of one day a model that is as accurate as possible for application to the coming day.

The general form of the empirical expressions presented in Chapter 2, including Equations (2.4), (2.6), and (2.8), is as follows:

$$Yp(k) = \sum_{n=0}^{N-1} H(n, k)X(n, k) \ , \qquad (3.1)$$

where $Yp(k)$ is the predicted value of the dependent variable at time k; $H(0, k) = 1$; $H(n, k)$ are the values or functions of accessible system variables at time k; and $X(n, k)$ represents the "state" of the empirical expression at time k. If these constants are determined by regression, $X(n, k)$ does not vary with time k until such time as a new regression is carried out yielding new regression coefficients and, thus, a new state of the expression. Since the determination of the constants X in one empirical expression is independent of the determination of the constants X in any other, the approach to maximizing overall model accuracy will be to maximize the accuracy of each individual empirical expression. For the present purpose, the *state of the model* will refer to the set of coefficients, X, for all empirical expressions, as distinct from the *state of the system,* which includes tank depths and various flows and pressures.

A distinction must be made between modeling and measurement errors. The actual value of the dependent variable is related to the predicted value as follows:

$$Ya(k) = Yp(k) + Np(k), \tag{3.2}$$

where $Np(k)$ is the error at time k due to modeling errors. Such errors arise from inaccurate estimation of the empirical equation state constants $X(n, k)$, as well as from the failure to include in the expression all independent variables that affect the dependent variable. One such unincluded variable that affects all of the empirical expressions is the actual water consumption by every customer, which obviously cannot be accounted for.

The actual value of the dependent variable can never be known exactly because of measurement error; this can be expressed by

$$Z(k) = Ya(k) + Nm(k), \tag{3.3}$$

where $Z(k)$ is the measured value of the dependent variable, which differs from the actual value by a measurement error term $Nm(k)$. Combining the last two equations, we obtain

$$Z(k) = Yp(k) + Nm(k) + Np(k). \tag{3.4}$$

For the present purpose, the two error terms are combined, and their sum will be called the measurement noise:

$$V(k) = Nm(k) + Np(k). \tag{3.5}$$

Finally, Equations (3.1), (3.4), and (3.5) are combined to obtain

$$Z(k) = \sum_{n=0}^{N-1} H(n, k)X(n, k) + V(k). \tag{3.6}$$

In Equation (3.6) the empirical expressions developed in the previous chapter have been written in terminology consistent with the Kalman filter equations that follow.

Regression or the Kalman Filter?

For the purpose of calculating unknown coefficients in the empirical expressions, data can be analyzed either in a batch mode or sequentially. Day long batches of data were used in the regressions in the last chapter. It is, of course, possible, at the end of each day, to analyze the past day's data in order to produce a macroscopic model for the next day. An advantage of this procedure is that no model need by hypothesized to account for the changing conditions. However, a macroscopic model based on one set of data does not take advantage of the analysis of previous days. As a result, abnormal data caused by emergency conditions will lead to a nonrepresentative model for the next day. It is possible to analyze the data of many days at one time. This would, however, involve storage and computation of large amounts of data, and the rationale for determining the number of days to include in the regression is not apparent.

The analogy between water distribution and power distribution suggests that some power system modeling procedures might be applicable. In 1970, Larson, Tinney, and Peschon [43] examined the problem of determining the state of a power system. They used weighted least squares, a method of which unweighted regression is a special case. Appendix A describes regression, where all independent variables carry equal weight. In weighted least squares, each independent variable may be assigned a weight that determines the influence of that independent variable on the dependent variable. Frequently, weighting is determined by the inverse of the independent variable covariance matrix, with the result that the more erratic independent variables are weighted less than the steadier ones.

While the current problem is to find the state of the model, the work of Larson *et al.* is an analogous search for the state of the system. Both the weighted least-squares and the regression methods are static estimators, and both are a special cases of the Kalman filter [77]. Debs and Larson [17] extended this power system work to form a dynamic estimator of the system state, using the Kalman filter. In taking this step, state estimates are a function of both current and past measurements. The Kalman filter has several properties suitable for the special requirement of making a model available at the end of one day that is accurate for the next. In particular, by the adjustment of several filter parameters, the filter "bandwidth" can be varied to attain the desired filter reaction to change. As a computational convenience, when the filter carries out regression recursively, no matrix inversion is required, and the computer storage needed is less than that required for batch regression. Finally, the filter takes advantage of data from past days, automatically narrowing its bandwidth as particularly noisy abnormal data are processed.

Design of the Kalman Filter

This section presents an algorithm for keeping the macroscopic model current by recursive regression carried out by a Kalman filter. The Kalman filter equations are presented first, using the notation of Sorensen [76], whose work contains a good intuitive development of the filter equations.

Next, after modifications are made to carry out regression recursively, the state and measurement noises are set. Finally, a numerical example demonstrates the superiority of the Kalman filter approach to batch processing of data by regression. The comparison covers a period of one week.

The state vector, $\mathbf{X}(k)$, at time k, is modeled by a relation involving state $\mathbf{X}(k-1)$ and the state transition matrix $\boldsymbol{\phi}(k, k-1)$, with additive white noise:

$$\mathbf{X}(k) = \boldsymbol{\phi}(k, k-1)\mathbf{X}(k-1) + \mathbf{W}(k-1), \quad k = 1, 2, \ldots . \tag{3.7}$$

Measurements $\mathbf{Z}(k)$ are related linearly to the state vector by a matrix $\mathbf{H}(k)$, with additive white noise:

$$\mathbf{Z}(k) = \mathbf{H}(k)\mathbf{X}(k) + \mathbf{V}(k) . \tag{3.8}$$

The properties of the zero-mean white-noise terms are summarized by the following expected values:

$$E[\mathbf{V}(k)] = \mathbf{0} , \tag{3.9}$$

$$E[\mathbf{V}(k)\mathbf{V}(j)^T] = \mathbf{R}(k)\delta(k, j) , \tag{3.10}$$

$$E[\mathbf{V}(k)\mathbf{W}(j)^T] = \mathbf{0} , \tag{3.11}$$

$$E[\mathbf{W}(k)] = \mathbf{0} , \tag{3.12}$$

$$E[\mathbf{W}(k)\mathbf{W}(j)^T] = \mathbf{Q}(k)\delta(k, j) , \tag{3.13}$$

where $\mathbf{R}(k)$ is the measurement noise covariance matrix; $\mathbf{Q}(k)$, the state noise covariance matrix; $\delta(k, j) = 1$ for $k = j$; and $\delta(k, j) = 0$ for $k \neq j$. The estimate $\hat{\mathbf{X}}(k)$ of $\mathbf{X}(k)$ is made such that the following (a least-squares expression) is minimized:

$$E[\{\hat{\mathbf{X}}(k) - \mathbf{X}(k)\}^T \{\hat{\mathbf{X}}(k) - \mathbf{X}(k)\}] . \tag{3.14}$$

The error covariance matrix is given by

$$\mathbf{P}(k) = E[\{\hat{\mathbf{X}}(k) - \mathbf{X}(k)\} \{\hat{\mathbf{X}}(k) - \mathbf{X}(k)\}^T]. \tag{3.15}$$

Therefore, the estimation optimality criterion of Equation (3.14) is equivalent to

$$\text{trace } [\mathbf{P}(k)] = \text{minimum} . \tag{3.16}$$

The trace operation calculates the sum of the main diagonal terms of the matrix argument. The estimation equation is as follows:

$$\begin{aligned} \hat{\mathbf{X}}(k) = {} & \boldsymbol{\phi}(k, k-1)\hat{\mathbf{X}}(k-1) \\ & + \mathbf{K}(k)[\mathbf{Z}(k) - \mathbf{H}(k)\boldsymbol{\phi}(k, k-1)\hat{\mathbf{X}}(k-1)]. \end{aligned} \tag{3.17}$$

The current estimate, $\hat{\mathbf{X}}(k)$, is equal to the time propagation or projection of the past estimate, $\boldsymbol{\phi}(k, k-1)\hat{\mathbf{X}}(k-1)$, plus a matrix, $\mathbf{K}(k)$, multiplied by the difference between the measured value $\mathbf{Z}(k)$ and the projected measured value, $\mathbf{H}(k)\boldsymbol{\phi}(k, k-1)\hat{\mathbf{X}}(k-1)$. Thus, the estimates $\hat{\mathbf{X}}(k)$ are linear combinations of the measured dependent variables $[\mathbf{Z}(k), \ldots, \mathbf{Z}(1)]$ and the independent variables $[\mathbf{H}(k), \ldots, \mathbf{H}(1)]$. Define $\mathbf{P}(k)'$ to be related to the covariance matrix and state noise, as follows:

$$\mathbf{P}(k)' = \boldsymbol{\phi}(k, k-1)\mathbf{P}(k-1)\boldsymbol{\phi}(k, k-1)^T + \mathbf{Q}(k-1). \qquad (3.18)$$

The sequence of gains, $\mathbf{K}(k)$, is then evaluated using this equation and Equations (3.19) and (3.20):

$$\mathbf{K}(k) = \mathbf{P}(k)'\mathbf{H}(k)^T[\mathbf{H}(k)\mathbf{P}(k)'\mathbf{H}(k)^T + \mathbf{R}(k)]^{-1}, \qquad (3.19)$$

where

$$\mathbf{P}(k) = \mathbf{P}(k)' - \mathbf{K}(k)\mathbf{H}(k)\mathbf{P}(k)'. \qquad (3.20)$$

In summary, the state $\mathbf{X}$ is assumed to be described by the dynamic equation [Equation (3.7)], where the error term $\mathbf{W}$ accounts for deficiencies in the assumed state transition matrix $\boldsymbol{\phi}$. Measurements are made according to Equation (3.8), with measurement error accountable by the $\mathbf{V}$ terms. The current estimate $\hat{\mathbf{X}}(k)$ is made in terms of the past estimate $\hat{\mathbf{X}}(k-1)$ and the present measurement $\mathbf{Z}(k)$. The optimal gain sequence $[\mathbf{K}(k), \ldots, \mathbf{K}(1)]$ is derived in such a way as to minimize the sum square error in Equation (3.16). In most respects, the Kalman filter is the same as the Gauss least-squares estimation, with the important exception that the Kalman filter allows for dynamics through the state transition matrix $\boldsymbol{\phi}$ and its associated error $\mathbf{W}$.

In dealing with the empirical expressions one at a time, we relate a single dependent variable to several independent variables, as shown in Equation (3.1). Therefore, the measurement, $Z(k)$, is related to the predicted independent variable value by an error term, $V(k)$, as follows:

$$Z(k) = \sum_{n=0}^{N-1} H(n, k)X(n, k) + V(k). \qquad (3.21)$$

If the independent variable measurement matrix is defined as a row vector,

$$\mathbf{H}(k) = [1 \quad \mathbf{H}(1, k) \quad H(2, k) \quad \ldots \quad H(N-1, k)], \qquad (3.22)$$

and the vector of parameters to be estimated is defined as

$$\mathbf{X}(k) = [X(0, k) \quad X(1, k) \quad \ldots \quad X(N-1, k)]^T, \qquad (3.23)$$

then the matrix equivalent of Equation (3.21) is as follows:

$$Z(k) = \mathbf{H}(k)\mathbf{X}(k) + V(k). \qquad (3.24)$$

This is the same as the measurement equation [Equation (3.8)], except that $Z(k)$ and $V(k)$ are not vectors. It follows then that $R(k)$ is a 1×1 matrix, the measurement noise variance. To summarize, $Z(k)$ is the measured dependent variable: $\mathbf{H}(k)$ contains the independent variable measurements, and $\mathbf{X}(k)$ is the state of the empirical relation, described in the previous chapter by constants determined from a regression analysis of batches of operation data.

While the state of the expressions is expected to vary dynamically because of such factors as pipe aging, such variation on a daily basis is quite small; in addition, it is difficult to assign any dynamics to these state changes. These variations can be taken into account through an additive noise term. Although in some cases the noise might be nonwhite, thus requiring appropriate modifications of the Kalman filter equations, for the present purpose white noise is assumed. Thus, we set the state transition matrix equal to unity and Equation (3.7) is modified as follows:

$$\mathbf{X}(k) = \mathbf{X}(k-1) + \mathbf{W}(k-1). \tag{3.25}$$

Furthermore, state and measurement noise will be considered to be time stationary. With these modifications in mind, the Kalman filter equations are rewritten. The matrix $\mathbf{P}(k)'$ has been absorbed into the other equations:

$$\mathbf{X}(k) = \mathbf{X}(k-1) + \mathbf{W}(k-1), \tag{3.26}$$

$$Z(k) = \mathbf{H}(k)\mathbf{X}(k) + V(k), \tag{3.27}$$

$$\hat{\mathbf{X}}(k) = \hat{\mathbf{X}}(k-1) + \mathbf{K}(k)[Z(k) - \mathbf{H}(k)\hat{\mathbf{X}}(k-1)], \tag{3.28}$$

$$\mathbf{P}(k) = [\mathbf{I} - \mathbf{K}(k)\mathbf{H}(k)]\,[\mathbf{P}(k-1) + \mathbf{Q}], \tag{3.29}$$

$$\mathbf{K}(k) = [\mathbf{P}(k-1) + \mathbf{Q}]\mathbf{H}(k)^T\{\mathbf{H}(k)\,[\mathbf{P}(k-1) + \mathbf{Q}]\mathbf{H}(k)^T + R\}^{-1}. \tag{3.30}$$

Before these equations are applied, a few points should be made:

1. The term in braces ({ . . . }) in Equation (3.30) is a scalar, so the matrix inversion is reduced to division.
2. It is apparent in Equation (3.30) that the optimal gain sequence $\mathbf{K}(k)$ will be unaffected if $\mathbf{P}(k-1)$, $\mathbf{Q}$, and R are multiplied by the same factor.
3. As $\mathbf{K}(k)$ is increased in Equation (3.28), the estimate $\hat{\mathbf{X}}(k)$ will react more rapidly to the difference between the dependent variable measurement $Z(k)$ and the expected measurement $\mathbf{H}(k)\hat{\mathbf{X}}(k-1)$. This amounts to an increase in bandwidth.
4. From Equations (3.29) and (3.30), $\mathbf{K}(k)$, (and therefore bandwidth) increases as the state noise $\mathbf{Q}$ increases relative to the measurement noise R.

In the discussion that follows, it should be remembered that the Kalman filter is not being applied in a completely conventional way; rather it is being adapted to accomplish a specific goal. To keep the empirical model current, each different empirical expression will have its own Kalman filter, tailored to process the specific data of each expression. The filters will be "tuned," based on the analysis of a few, say ND, days. If it is assumed that the noise characteristics are approximately constant [$Q(k) = Q$ and $R(k) = R$], the filters will then be usable far into the future. Each of the ND days contains NT data points, consisting of the dependent variable $Z(k)$ and the $N - 1$ independent variables in $\mathbf{H}(k)$, for $1 \leqq k \leqq (NT \cdot ND)$. While the state estimates $\hat{\mathbf{X}}(k)$ are available at the end of every time step k, the ones at the end of each day are of special interest. Define ${}^{\delta}\hat{\mathbf{X}}$ as the state vector as it exists after the end of day δ. The Kalman filter will be considered to be optimally tuned by the relative magnitude of $\mathbf{Q}$ and R when the following mean-square-error objective function is minimized:

$$J = \frac{1}{ND \cdot NT} \sum_{nd=2}^{ND} \sum_{nt=1}^{NT} \left[Z(k) - \mathbf{H}(k)\, {}^{\delta}\hat{\mathbf{X}} \right]^2 , \tag{3.31}$$

where $\delta = nd - 1$ and $k = (nd - 1)NT + nt$. The term within the brackets is the difference between a measured value of the dependent variable and the predicted value for that variable, based on values of the independent variables taken at the same time and applied to the empirical expression using states as they existed at the end of the previous day.

In order to begin the recursive computation, initial values of R, $\mathbf{Q}$, and $\mathbf{P}$ must be set. To begin with maximum uncertainty, the initial guess of the parameter vector $\hat{\mathbf{X}}(0)$ will be set equal to the zero vector $\mathbf{0}$. The scalar measurement variance Ro is initially set to a large value, which will later be scaled to a more realistic value.

The Kalman filter is tuned on the basis of several days of data; once tuned, it will be used far into the future. In order to help in setting the state noise, it is necessary to perform batch mode regressions on the base ND days. The vector ${}^{\delta}\tilde{\mathbf{X}}$ is the vector of regression parameters for δth day. The initial state covariance matrix can be set equal to the sample covariance matrix, $\mathbf{Q}o$, whose terms are as follows:

$$\mathbf{Q}o(i, j) = \frac{1}{ND - 1} \sum_{\delta=1}^{ND} [({}^{\delta}\tilde{X}(i) - \bar{X}(i))\,({}^{\delta}\tilde{X}(j) - \bar{X}(j)] , \tag{3.32}$$

where ${}^{\delta}\tilde{X}(i)$ is the ith member of ${}^{\delta}\tilde{\mathbf{X}}$, and $\bar{X}(i)$ is the average of the ith member over ND days. Finally, it can be shown [76] that Equation (3.29) can be written as follows:

$$\mathbf{P}^{-1}(k) = [\mathbf{P}(k - 1)]^{-1} + \mathbf{H}^T(k)R^{-1}\mathbf{H}(k) . \tag{3.33}$$

Again, for maximum uncertainty, if $\mathbf{P}(0) = \infty$, then

$$\mathbf{P}(1) = [\mathbf{H}(1)^T R^{-1}\mathbf{H}(1)]^{-1} . \tag{3.34}$$

While the matrix $\mathbf{Q} = \mathbf{Q}o$ is used on the first pass through the set of data points covering *ND* days, for subsequent passes the matrix is modified by a constant c:

$$\mathbf{Q} = c\mathbf{Q}o\,. \tag{3.35}$$

Thus, by varying the constant c while holding the measurement noise R constant, the filter bandwidth is varied. The best value of c results from a unimodal search with the value J in Equation (3.31) being the objective function to be minimized. Intuitively, the existence of an optimum bandwidth can be argued as follows: If the bandwidth is too narrow, the filter will not react to changing conditions over the *ND* days, thus causing error to accumulate in the latter days. On the other hand, if the bandwidth is too wide, the model state as it exists at the end of the day will be largely the result of the last few measurements and not necessarily representative of the entire day. Thus an optimum bandwidth would be expected to exist somewhere in between as indeed it does in the numerical examples. This approach of selecting the state covariance matrix is essentially that presented by Debs and Larson [17] with the exception that they do not use off-diagonal terms in $\mathbf{Q}o$. It was found numerically that the inclusion of these terms increased accuracy.

Once the optimum $\mathbf{Q} = c\mathbf{Q}o$—which leads to minimum J—has been found, the values R, $\mathbf{Q}$, and $\mathbf{P}$ may be scaled by the ratio (J/Ro) to obtain more reasonable values. As it was pointed out previously, this scaling does not affect the filter performance. The objective function J, is essentially the estimate of R with the exception that, in the place of the unknown $\mathbf{X}(k)$ at every time interval, the daily parameter vector ${}^{\delta}\hat{\mathbf{X}}$ is used.

The special purpose Kalman filter tuning can be summarized in the following steps:

1. A set of operating data covering several days is obtained.
2. Each empirical expression has its own filter, so one expression is selected.
3. Values of the dependent variable are set equal to the measurement array.
4. Values of the independent variables are set equal to the H matrix array.
5. R, $\mathbf{P}$, and $\mathbf{Q}$ are initialized.
6. While R is held constant, the objective function J [Equation (3.31)] is minimized by scaling the matrix $\mathbf{Q}$. A single value of J results from a complete pass through all of the data using the filtering equations [Equations (3.26) through (3.30)]. Prior to each pass, the filter is run to an approximate steady state on the data of one day. A true equilibrium is not possible because the time varying $\mathbf{H}(k)$ causes the filter to be nonstationary.

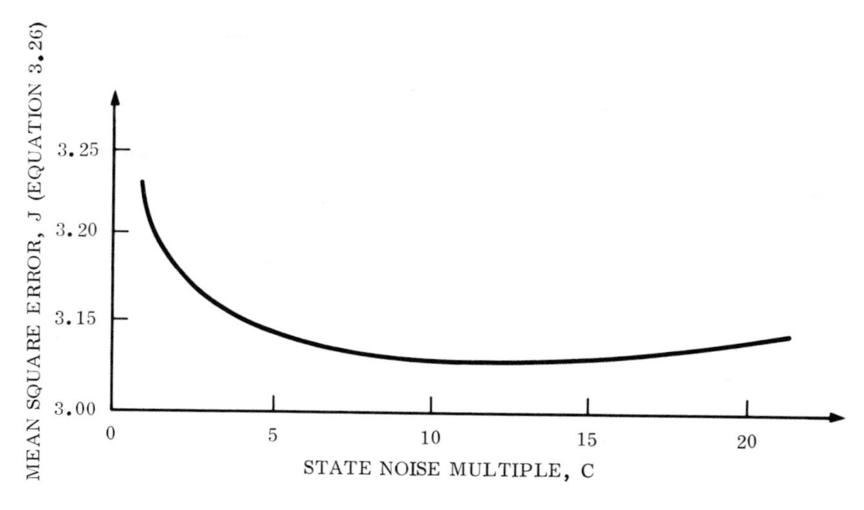

Figure 3-1. Mean-Square Error Versus State Noise Multiple for Head Drop from Pump Station 1 to Tank 1

7. R, $\mathbf{P}$, and $\mathbf{Q}$ are scaled by the same factor to obtain reasonable values.

Numerical Example

Data were obtained for the week of 10 January through 16 January 1972 from the Torresdale High Service-Fox Chase Booster District. Regression analyses were carried out for each of the first six days; then the empirical model as it existed at the end of each day, using regression results, was used in a validation using the data for the next day as the basis of comparison. In the same way, the data of the first six days were processed by the Kalman filter, and the resulting empirical models validated with the following day's data. Thus a direct comparison of accuracy was made.

The single figure of merit of a validation of an empirical model is the modified multiple correlation coefficient as defined in the previous chapter [Equations (2.41) through (2.46)]. In this special definition, the total sum-square error, SST, is identical from one day to the next. For this reason, the overall multiple correlation coefficient for a period of L days, ρ_L, can be written in terms of the individual daily multiple correlation coefficients:

$$\rho_L = \left(\sum_{\delta=1}^{L} \rho_\delta^2 / L \right)^{1/2}. \tag{3.36}$$

The Kalman filter tuning was carried out as described at the end of the previous section. Figure 3-1 shows the results of the search for an optimum state noise multiple c, found in Equation (3.35), leading to the minimum

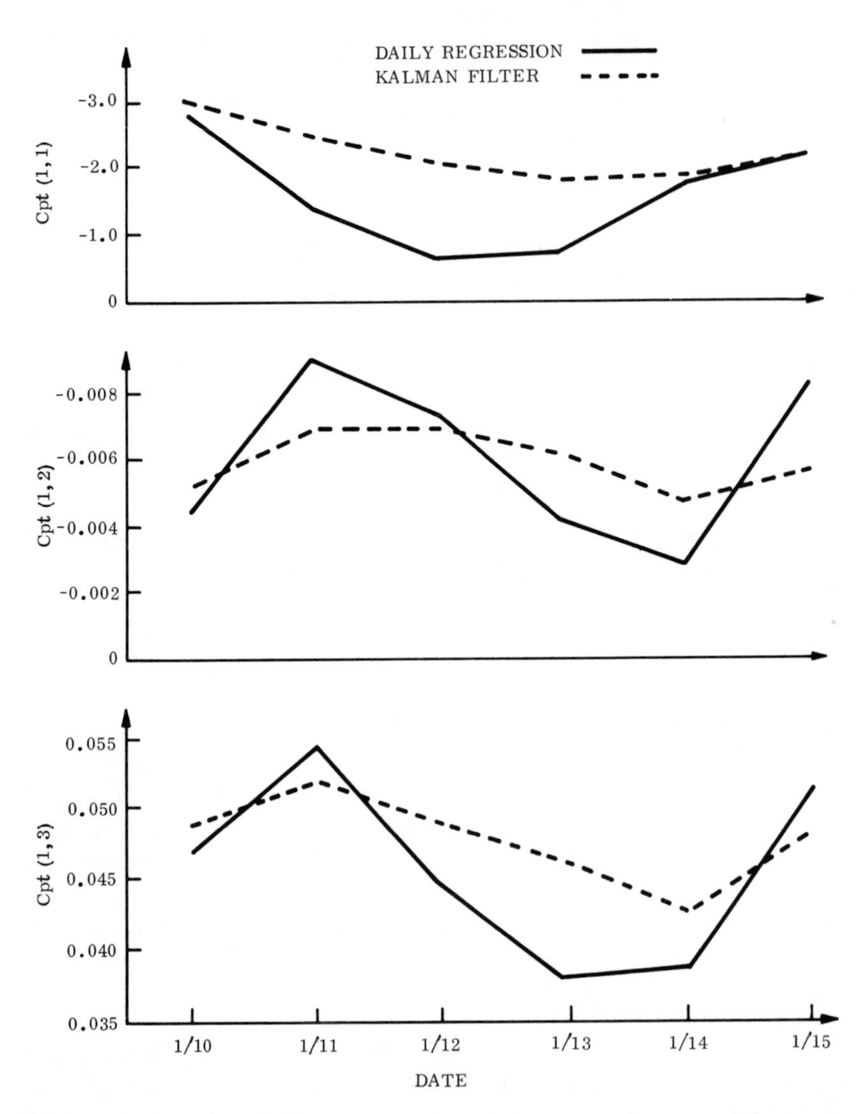

Figure 3-2. Variations in Parameters: Head Drop from Pump 1 to Tank 1

mean square error objective function [Equation (3.31)]. As Figure 3-1 shows, the value of c is not at all critical. For this reason, rather than set up a formal computerized search, the optimum value of c was found by an informal dichotomous search using a time-sharing terminal.

For the same empirical expression (head drop from pump station 1 to tank 1), the equation coefficients are plotted in Figure 3-2 as they existed at the end of the first six days of the week. As expected, those coefficients

Table 3-1
Daily Empirical Expression Coefficients for 10-15 January 1972 for Head Drop from Pump Station 1 to Tank 1:

$$Hpt(1) = Cpt(1, 1) + Cpt(1, 2)Qd^{1.85} + Cpt(1, 3)Qp(1)^{1.85}$$

	1/10	1/11	1/12	1/13	1/14	1/15
Regression						
$Cpt(1, 1)$	−2.89	−1.44	−0.66	−0.72	−1.79	−2.19
$Cpt(1, 2)$	−0.00412	−0.00902	−0.00717	−0.00399	−0.00275	−0.00824
$Cpt(1, 3)$	0.0476	0.0555	0.0449	0.0384	0.0389	0.0525
Kalman filter						
$Cpt(1, 1)$	−3.02	−2.51	−2.11	−1.82	−1.85	−2.16
$Cpt(1, 2)$	−0.0050	−0.0068	−0.0069	−0.0060	−0.0046	−0.0067
$Cpt(1, 3)$	0.0494	0.0530	0.0500	0.0468	0.0434	0.0489

Table 3-2
Daily Empirical Expression Coefficients for 10-15 January 1972 for Head Drop from Pump Station 2 to Tank 2:

$$Hpt(2) = Cpt(2, 1) + Cpt(2, 2)Qd^{1.85} + Cpt(2, 3)Qp(1)^{1.85} + Cpt(2, 4)Qp(2)^{1.85}$$

	1/10	1/11	1/12	1/13	1/14	1/15
Regression						
$Cpt(2, 1)$	−2.79	−1.68	4.03	1.65	0.085	−1.78
$Cpt(2, 2)$	−0.0134	−0.0136	−0.0114	−0.0098	−0.0091	−0.0131
$Cpt(2, 3)$	0.0013	−0.0001	−0.0184	−0.0133	−0.0136	−0.0001
$Cpt(2, 4)$	0.301	0.304	0.284	0.282	0.296	0.290
Kalman filter						
$Cpt(2, 1)$	−3.34	−1.93	0.801	1.36	0.867	−1.89
$Cpt(2, 2)$	−0.0138	−0.0124	−0.0233	−0.0133	−0.0107	−0.0130
$Cpt(2, 3)$	0.0033	−0.0025	−0.0090	−0.0116	−0.0015	−0.0002
$Cpt(2, 4)$	0.301	0.304	0.292	0.290	0.291	0.292

resulting from the Kalman filter show less variation than those computed from daily regressions. Daily empirical expression coefficients from the Kalman filter and regressions are tabulated in Tables 3-1 through 3-5 for all of the empirical expressions. Multiple regression analyses and stepwise regression analyses previously eliminated the unnecessary flow into tank 2, as well as several insignificant independent variables.

Table 3-3
Daily Empirical Expression Coefficients for 10-15 January 1972 for Flow into Tank 1:

$$Qt(1, 1) = Ct(1, 1) + Ct(1, 2)Qd + Ct(1, 3)Qp(1) + Ct(1, 4)Qp(2) + Ct(1, 5)Ht(1)^{1/1.85} + Ct(1, 6)Ht(2)^{1/1.85}$$

	1/10	*1/11*	*1/12*	*1/13*	*1/14*	*1/15*
Regression						
$Ct(1, 1)$	−31.82	15.97	−3.97	−49.1	0.23	−30.8
$Ct(1, 2)$	−0.752	−0.721	−0.757	−0.765	−0.754	−0.760
$Ct(1, 3)$	−4.96	−3.54	−7.91	−3.84	−6.81	−5.06
$Ct(1, 4)$	6.41	2.92	8.16	5.98	6.83	6.45
$Ct(1, 5)$	0.911	0.852	0.903	0.946	0.936	0.966
$Ct(1, 6)$	0.331	0.291	0.266	0.421	0.344	0.296
Kalman filter						
$Ct(1, 1)$	−32.24	−18.44	−15.73	−18.67	−13.34	−22.88
$Ct(1, 2)$	0.753	0.741	−0.745	−0.748	−0.740	−0.754
$Ct(1, 3)$	−5.17	−4.12	−5.29	−5.15	−3.66	−5.74
$Ct(1, 4)$	6.64	4.97	6.03	6.02	4.28	6.81
$Ct(1, 5)$	0.911	0.900	0.904	0.905	0.901	0.918
$Ct(1, 6)$	0.331	0.320	0.319	0.326	0.322	0.333

Table 3-4
Daily Empirical Expression Coefficients for 10-15 January 1972 for Pressure Point 1:

$$Pn(1) = Cn(1, 1) + Cn(1, 2)Qd^{1.85} + Cn(1, 3)Hd(1) + Cn(1, 4)Hd(2) + Cn(1, 8)Ht(1)$$

	1/10	*1/11*	*1/12*	*1/13*	*1/14*	*1/15*
Regression						
$Cn(1, 1)$	−121.5	−105.4	−107.5	−81.2	−111.8	−115.2
$Cn(1, 2)$	−0.0016	−0.0020	−0.0024	−0.0032	−0.0021	−0.0022
$Cn(1, 3)$	0.116	0.168	0.138	0.193	0.079	0.177
$Cn(1, 4)$	0.0257	0.0215	0.0343	0.0342	0.0082	0.0133
$Cn(1, 8)$	0.448	0.351	0.376	0.241	0.477	0.381
Kalman filter						
$Cn(1, 1)$	−121.0	−116.7	−112.5	−99.2	−107.0	−106.9
$Cn(1, 2)$	−0.0016	−0.0017	−0.0020	−0.0024	0.0021	−0.0022
$Cn(1, 3)$	0.120	0.134	0.140	0.148	0.130	0.142
$Cn(1, 4)$	0.0260	0.0271	0.0268	0.0340	0.0265	0.0254
$Cn(1, 8)$	0.443	0.415	0.398	0.340	0.391	0.380

Table 3-5
Daily Empirical Expression Coefficients for 10-15 January 1972 for Pressure Point 2:

$$Pn(2) = Cn(2, 1) + Cn(2, 2)Qd^{1.85} + Cn(2, 3)Hp(1) + Cn(2, 4)Hp(2) + Cn(2, 8)Ht(1) + Cn(2, 9)Ht(2)$$

	1/10	*1/11*	*1/12*	*1/13*	*1/14*	*1/15*
Regression						
$Cn(2, 1)$	−138.9	−120.7	−121.1	−118.5	−132.0	−130.2
$Cn(2, 2)$	−0.00321	−0.00272	−0.00310	−0.00239	−0.00347	−0.00309
$Cn(2, 3)$	0.1060	0.0596	0.0677	0.0257	0.0377	0.0864
$Cn(2, 4)$	0.0572	0.0737	0.0761	0.0593	0.0606	0.0535
$Cn(2, 8)$	0.387	0.230	0.197	0.205	0.377	0.356
$Cn(2, 9)$	0.018	0.150	0.174	0.217	0.073	0.046
Kalman filter						
$Cn(2, 1)$	−138.9	−130.4	−123.9	−123.5	−127.7	−127.7
$Cn(2, 2)$	−0.0032	−0.0030	−0.0029	−0.0029	−0.0030	−0.0031
$Cn(2, 3)$	0.109	0.104	0.099	0.090	0.066	0.080
$Cn(2, 4)$	0.0572	0.0622	0.0673	0.0667	0.0635	0.0630
$Cn(2, 8)$	0.388	0.297	0.223	0.221	0.163	0.165
$Cn(2, 9)$	0.014	0.079	0.134	0.144	0.142	0.126

Finally, Table 3-6 shows the multiple correlation coefficients resulting from the model validations, with the Kalman filter showing a superior result. In addition to the improvement in accuracy, it should be kept in mind that the Kalman filter offers computational advantages over daily regressions, including no need for matrix inversion and reduced computer storage space.

Further Application

Centrifugal pumps presumably have a fixed head-flow relationship such as the curve shown in Figure 2-2. In practice, however, it has been found that logged data can show a considerable point scatter, thus requiring the pump curve to be found by statistical analysis. Pump curves found by regression, which worked well with data from one season of the year, caused serious errors when applied to another season. Explanations for point scatter and drift have varied from pump wear to telemetry noise and even to the changing of pump impellors [7, 33, 59, 69]. Whatever the reason, the continuous Kalman filter monitoring of pump head-flow relationships seems desirable. The form of the pump head-flow relation was given in Equation (2.3). This equation is a special case of Equation (3.1), so the

Table 3-6
Multiple Correlation Coefficients Resulting from Daily Model Validations for 11-16 January 1972

	1/11	*1/12*	*1/13*	*1/14*	*1/15*	*1/16*	*Overall*
I Regression coefficients from previous day:	0.949	0.950	0.928	0.920	0.942	0.917	0.934
II Kalman filter coefficients from previous day:	0.944	0.956	0.947	0.934	0.930	0.930	0.940

adapting of pump equations would follow exactly the same way as was described for the empirical expressions.

4 Time Varying Demand Study

For the purpose of controlling water distribution by dynamic programming, the expected demand (mean value function) is required for at least one day into the future. The mean value function contains only the smooth, predictable time variations in demand. It is also necessary to determine the statistics of the unpredictable random variations that may be generated and added to the mean value function to form a realistic simulated demand function. This simulated actual demand is required in the simulation of a distribution system for the purpose of evaluating a control system.

Since relatively short-term predictions are needed, periodicities —rather than trends—in demand will be detected. Periodicities are extracted from demand data by Fourier analysis. Statistical tests are performed to determine the number of harmonics that are significant in describing time varying demand as well as which days of the week lead to mean value functions significantly different from those of other days. Initially, winter data are analyzed since these are relatively free of weather effects. Once this is done, summer data are analyzed, showing the effects of rain and temperature on total demand.

Existing Demand Studies

Most demand studies are concerned with discovering those factors that affect consumption. Typical contributing factors include the following [62, 87]:

1. Market value of residences.
2. Population density.
3. Housing density.
4. Water price.
5. Climate.

Using such factors as these, long-term projections are made for the purpose of planning and design. The minimum time period considered in such studies is seldom less than one month. Intervals down to one day have been considered in the study of the relationship between consumption due to lawn sprinkling and water pricing [30, 44].

Some authors are more interested in demand data from a time series [4] standpoint. Gracie [27, 1966] used harmonic and spectral analysis to discover periodicities in daily and weekly consumption, as well as analysis of variance to determine which days of the week have significantly different average consumptions. Similarly, in a very detailed work, Salas-La Cruz and Yevjevich [72, 1972] used harmonic analysis to identify and estimate periodicities.

While all of the above references contain some aspects applicable to the analysis of time varying demand data suitable for the current requirement, none are completely adequate. For this reason, this chapter presents a straightforward application of Fourier and statistical analysis, including numerical results for the specific district studied.

Fourier Series Analysis

Before statistical tests are set up, a few Fourier equations will be summarized. A time varying demand for one day can be exactly represented by an infinite series:

$$Qd(t) = a1(0) + \sum_{j=1}^{\infty} a1(j)\cos(2\pi jt) + \sum_{j=1}^{\infty} b1(j)\sin(2\pi jt), \tag{4.1}$$

where t represents the fraction of one day ($0 \leqq t \leqq 1$); and $a1(j)$, $b1(j)$ are true harmonic coefficients. If, over the period of one day, NT periodic samples $[q(n)]$ are made of $Qd(t)$, the true function can be approximated by the following:

$$\widehat{Qd}(t;NH) = a2(0) + \sum_{j=1}^{NH} a2(j)\cos(2\pi jt) + \sum_{j=1}^{NH} b2(j)\sin(2\pi jt), \tag{4.2}$$

where NH is the number of harmonics; $a2(j)$, the estimate of $a1(j)$; and $b2(j)$, the estimate of $b1(j)$. The difference between actual and estimated demand is the error:

$$e(t) = Qd(t) - \widehat{Qd}(t;NH). \tag{4.3}$$

The least-squares estimates of $a2(j)$ and $b2(j)$ are as follows:

$$a2(0) = \frac{1}{NT}\sum_{n=1}^{NT} q(n); \tag{4.4}$$

$$a2(j) = \frac{2}{NT}\sum_{n=1}^{NT} q(n)\cos\left(\frac{2\pi jn}{NT}\right), \quad 1 \leqq j \leqq NH; \tag{4.5}$$

$$b2(j) = \frac{2}{NT}\sum_{n=1}^{NT} q(n) \sin\left(\frac{2\pi jn}{NT}\right), \qquad 1 \leqq j \leqq NH. \tag{4.6}$$

These estimates are mutually independent and unbiased [35]. The variances are as follows:

$$\text{Var}\,[a2(0)] = (1/NT)\,\text{Var}\,[e(t)]; \tag{4.7}$$

$$\text{Var}\,[a2(j)] = \text{Var}\,[b2(j)] = (2/NT)\,\text{Var}\,[e(t)], \quad t = n/NT, \quad j > 0. \tag{4.8}$$

Determination of Significant Harmonics

It is not obvious from the data what number of harmonics, NH, is sufficient to represent $Qd(t)$ by

$$\widehat{Qd}(t;\, NH).$$

No harm is caused by carrying insignificant harmonics in the mean value function of Equation (4.2), but, in the interest of eliminating unnecessary computation, they should be removed.

The sum-square error between actual and estimated data points remaining, when NH harmonics are used in

$$\widehat{Qd}(t;\, NH),$$

is as follows:

$$SSE(NH) = \sum_{n=1}^{NT}\left[\, q(n) - \widehat{Qd}\left(\frac{n}{NT};\, NH\right)\right]^2. \tag{4.9}$$

The sum-squared error "removed" by the addition of the kth harmonic [$a2(k)$ and $b2(k)$] is then [$SSE(k-1) - SSE(k)$]. To test the hypothesis that $a2(k) = b2(k) = 0$, the F ratio is formed:

$$F(k) = \frac{[SSE(k-1) - SSE(k)]/2}{[SSE(k)\;/\;(NT-2k-1)]}. \tag{4.10}$$

The significance level of the test is rather arbitrary. Suppose that we wish to retain only those harmonics whose coefficients are nonzero with 95 percent confidence. Then, if $F_{0.95}$ is the 95 percent cumulative value of the F distribution with 2 and $NT-2k-1$ degrees of freedom, and if

$$F(k) > F_{0.95}, \tag{4.11}$$

it may be said with 95 percent confidence that $a2(k)$ and $b2(k)$ are both significantly different from zero.

Six weeks of winter data were analyzed (see Figure 2-11 for a typical set of forty-eight half-hourly data points). First, all Mondays were analyzed, then all Tuesdays, and so on. If the variations apparent in the figure are considered, it is not surprising that the results were not the same for every day of the week. All weekdays showed the first six harmonics to be significant, with the addition of the tenth on Wednesday and Thursday. The weekend days showed only the first through third and fifth to be significant. Since the tenth on Wednesdays and Thursdays was just barely significant at the arbitrary 95 percent confidence, and since extra harmonics are not harmful, NH was assigned the value of six harmonics for all work to follow.

Significantly Different Days of the Week

Suppose that NW weeks of periodic demand data are available for all days of the week. If the harmonic coefficients were estimated by Equations (4.3), (4.4), and (4.5), and all those of the same day of the week were averaged, the result would be seven different mean value functions. It would not be apparent to what extent the difference is caused by different customer habits from one day to another or to what extent it is caused by random variations.

For the present purpose, the data point initially defined as $q(n)$ will be redefined; $q(i, n; k)$ is the demand at the nth time interval in the ith week of data for the kth day of the week, such that $1 \leqq n \leqq NT$, where n represents the sample number; $1 \leqq i \leqq NW$, where i represents the week number; and $1 \leqq k \leqq 7$, where k represents the day of the week. In keeping with this definition, the harmonic coefficient estimates of Equations (4.3), (4.4), and (4.5) are redefined to designate the week number and day of the week:

$$a2(i, 0; k) = \frac{1}{NT} \sum_{n=1}^{NT} q(i, n; k) ; \tag{4.12}$$

$$a2(i, j; k) = \frac{2}{NT} \sum_{n=1}^{NT} q(i, n; k) \cos\left(\frac{2\pi jn}{NT}\right), \quad 1 \leqq j \leqq NH ; \tag{4.13}$$

$$b2(i, j; k) = \frac{2}{NT} \sum_{n=1}^{NT} q(i, n; k) \sin\left(\frac{2\pi jn}{NT}\right), \quad 1 \leqq j \leqq NH . \tag{4.14}$$

The above are unbiased estimators of the respective true harmonic coefficients $a1(i, 0; k)$, $a1(i,j; k)$, and $b1(i,j; k)$. Rather than continue to write each of the three categories of harmonic coefficients, where possible, $g(i, j; k)$ will denote any of the true coefficients and $\hat{g}(i, j; k)$ their estimates.

If one considers that the random variations in total demand are caused by the cumulative effects of thousands of customers, it is not unreasonable to assume that demand random variations are approximately zero-mean Gaussian random variables. Since the harmonic coefficient estimates are linear combinations of these random variables, the estimates themselves must be Gaussian:

$$\hat{g}(i, j; k) = N[g(i, j; k), c(j)\sigma^2(i; k)]. \tag{4.15}$$

The above shown that $\hat{g}(i, j; k)$ is Gausian with mean $g(i, j; k)$ (showing unbiasedness) and variance $c(j)\sigma^2(i; k)$. $\sigma^2(i; k)$ is the variance of the values $q(i, n; k)$ about the mean-value function $\hat{q}(i, n; k)$ for day of the week k; $c(j) = 1/NT$ for $j = 0$; and $c(j) = 2/NT$ for $j > 0$. Recall that $\hat{g}(i, j; k)$ refers to either $a2(i, j; k)$ or $b2(i, j; k)$ for $j \geqq 1$ but only to $a2(i, j; k)$ for $j = 0$, because $b2(i, 0; k)$ does not exist. The true variance $\sigma^2(i; k)$ cannot be obtained from the data, but its unbiased estimate can:

$$\hat{\sigma}^2(i; k) = \frac{1}{(NT - 2NH - 1)} \sum_{n=1}^{NT} \left\{ q(i, n; k) - a2(i, 0; k) - \sum_{j=1}^{NH} \left[a2(i, j; k) \cos\left(\frac{2\pi jn}{NT}\right) + b2(i, j; k) \sin\left(\frac{2\pi jn}{NT}\right) \right] \right\}^2 \tag{4.16}$$

Therefore the random variable that follows is distributed according to Student's t distribution with $(NT - 2NH - 1)$ degrees of freedom:

$$\frac{\hat{g}(i, j; k) - g(i, j; k)}{[c(j)\hat{\sigma}^2(i; k)]^{1/2}} = t\,. \tag{4.17}$$

However, for practical purposes, the large number of degrees of freedom makes this essentially zero-mean unit-variance Gaussian.

In order to determine which days of the week are significantly different from one another, it is necessary to deal with harmonic coefficient estimates for each day k, averaged over the NW weeks of data:

$$\hat{\bar{g}}(j; k) = \frac{1}{NW} \sum_{i=1}^{NW} \hat{g}(i, j; k)\,. \tag{4.18}$$

The jth average coefficient in Equation (4.16) is exactly the same as the $(j \cdot NW)$th coefficient obtained by analyzing NW sets of data for the same day of the week placed end to end. In order to compare day $k1$ to $k2$, it is necessary to construct the confidence region of the following $(2NH + 1)$ random variables:

$$\hat{\bar{g}}(j; k1) - \hat{\bar{g}}(j; k2)\,. \tag{4.19}$$

The test is a multidimensional extension of the simple difference-of-means test. The confidence region is a $(2NH + 1)$-dimensional ellipsoid that, because of the independence of the estimates of the harmonic coefficients, contains no cross products.

To form the confidence region, it is necessary to find the variance of the average estimates in Equation (4.17). In order to find this variance, it is helpful to use the analogy of taking NW samples of NT objects each. Sum-square deviation is partitioned as follows:

$$\sum_{n=1}^{NT}\sum_{i=1}^{NW} [y(i, n) - \bar{\bar{y}}]^2 = NT\sum_{i=1}^{NW} [\bar{y}(i) - \bar{\bar{y}}]^2 + \sum_{n=1}^{NT}\sum_{i=1}^{NW} [y(i, n) - \bar{y}(i)]^2 , \tag{4.20}$$

where

$$\bar{\bar{y}} = \frac{1}{NT \cdot NW}\sum_{n=1}^{NT}\sum_{i=1}^{NW} y(i, n) ,$$

$$\bar{y}(i) = \frac{1}{NT}\sum_{n=1}^{NT} y(i, n) .$$

The overall variance of $y(i, n)$ on the left side of Equation (4.18) is equal to the sum of contributions of variations across the samples plus variations within the samples that are shown on the right side in the first and second terms, respectively.

The sample average $\bar{y}(i)$ is analogous to the harmonic coefficient $\hat{g}(i,j; k)$ obtained from the ith sample of NT points. The overall average

$$\bar{\bar{y}}$$

is analogous to the average coefficient

$$\hat{\bar{g}}(j; k)$$

taken over the NW samples. There is no direct analogy for $y(i, n)$ because the harmonic estimates cannot be obtained from each data point.

The first term in Equation (4.20) is then analogous to the following:

$$NT\sum_{i=1}^{NW} [\hat{g}(i, j; k) - \hat{\bar{g}}(j; k)]^2 . \tag{4.21}$$

If the second term is written as

$$\sum_{i=1}^{NW}\left\{\sum_{n=1}^{NT} [y(i, n) - \bar{y}(i)]^2\right\} , \tag{4.22}$$

the part in braces ({. . .}) contains the sum-square portion of the within sample variance. That variance has been shown to be equal to $c(j)\sigma^2(i; k)$, so from Equation (4.16) and this definition, the second term of Equation (4.20) is analogous to the following:

$$c(j)(NT - 2NH - 1)\sum_{i=1}^{NW} \hat{\sigma}^2(i; k). \qquad (4.23)$$

Finally, the overall sample variance of $y(i, n)$ is equal to Equation (4.20) divided by the number of degrees of freedom $(NT \cdot NW - 1)$. In the case of the harmonic analysis, there are only $(NT \cdot NW - 2NH - 1)$ degrees of freedom, so the sample variance of the harmonic coefficient estimates from Equations (4.20), (4.21), and (4.23) is as follows:

$$s^2(j; k) = \frac{1}{(NT \cdot NW - 2NH - 1)} \Bigg[NT \sum_{i=1}^{NW} [\hat{g}(i, j; k) - \hat{\bar{g}}(j; k)]^2$$

$$+ c(j)(NT - 2NH - 1) \sum_{i=1}^{NW} \sigma^2(i; k) \Bigg]. \qquad (4.24)$$

Generally, for two different days of the week, $k1$ and $k2$,

$$s^2(j; k1) \neq s^2(j; k2), \qquad (4.25)$$

but they are always close. If we use the approximation that they are nearly equal, the following random variable

$$[\hat{\bar{g}}(j; k1) - \hat{\bar{g}}(j; k2)] / [s^2(j; k1) + s^2(j; k2)]^{1/2} \qquad (4.26)$$

is approximately Gaussian with zero mean and unity variance.

The 100δ percent confidence region is a $(2NH + 1)$-dimensional ellipsoid centered at the origin. The semiaxes extend $X(\delta, NH)$ standard deviations in each direction, with $X(\delta, NH)$ defined implicitly as

$$\int_{-\infty}^{X(\delta, NH)} N(0, 1) = \frac{1}{2}[1 + \delta^{(2NH+1)-1}]. \qquad (4.27)$$

$N(0, 1)$ is the zero-mean unit-variance Gaussian density function. With respect to $a2$s and $b2$s, the 95 percent confidence ellipsoid is given for $EE = 1$:

$$\sum_{j=0}^{NH} \frac{[a^2(j; k1) - a^2(j; k2)]^2}{[s^2(j; k1) + s^2(j; k2)]X^2(0.95, NH)}$$

$$+ \sum_{j=1}^{NH} \frac{[b2(j; k1) - b2(j; k2)]^2}{[s^2(j; k1) + s^2(j; k2)]X^2(0.95, NH)} = EE. \qquad (4.28)$$

If, when this equation is evaluated, $EE > 1$, then it can be said with 95 percent confidence that the demand curve from day of the week $k1$ is significantly different than that of day $k2$.

The results of the analysis of $NW = 6$ weeks of data, using the Philadephia numerical example are as follows:

1. Comparison of any weekday with any other weekday: $EE < 1$.
2. Comparison with any weekday with either weekend day: $EE > 1$.

3. Comparison of the two weekend days: $EE < 1$.

The conclusion is that, for this set of data, it cannot be said with 95 percent confidence that there is a significant difference among weekdays or between weekend days. However, with 95 percent confidence, weekdays are different from weekend days. While this might seem like an obvious conclusion, it was not apparent from looking at the data that Monday or Friday would not stand alone as transitional days between weekend and weekdays.

Weekday harmonic coefficient estimates, all of which were not significantly different, were averaged to form a weekday mean value function; a mean value function was similarly formed for the weekend. The mean value functions are plotted in Figure 4-1. Note how the first peak in the weekend curve is delayed about three hours relative to the weekday peak, presumably because of less demanding personal weekend schedules.

Residual error—that is, the difference between data points and the mean value function, was analyzed. For both weekdays and weekend days, the residual error standard deviation was approximately $\hat{\sigma}_r = 4.2$ million gallons per day. The total daily consumption variance is related to the residual variance:

$$\hat{\sigma}^2_{\overline{Qd}} = \hat{\sigma}^2_r / NT. \tag{4.29}$$

In this particular case where $NT = 48$ points per day, the standard deviation of total daily consumption is $\hat{\sigma}_{\overline{Qd}} = 0.61$ million gallons out of a daily average of about 30 million gallons.

A very interesting result is that the residual error has no significant autocorrelation even at the minimum half-hour interval between demand measurements. This means that if, at one time, actual demand is above the expected value, there is no statistical basis for assuming that it will still be above a half-hour later. This result, coupled with the rather large residual error standard deviation, places into question the method of control described by McPherson [58] and quoted in Chapter 1. What is advocated are short-term projections based on current and expected conditions, as frequently as every quarter-hour. Such a control policy would be subject to excessive pump cycling. It is for this reason that the control concept described in Chapter 5 reacts to long-term trends (several hours in length) while remaining relatively insensitive to short-term variations.

Temperature and Rain Effects

During warm summer weather, average demand is greater than that experienced in cold weather; furthermore, demand has been observed to vary with both temperature and the presence of rain. Typical high-temperature

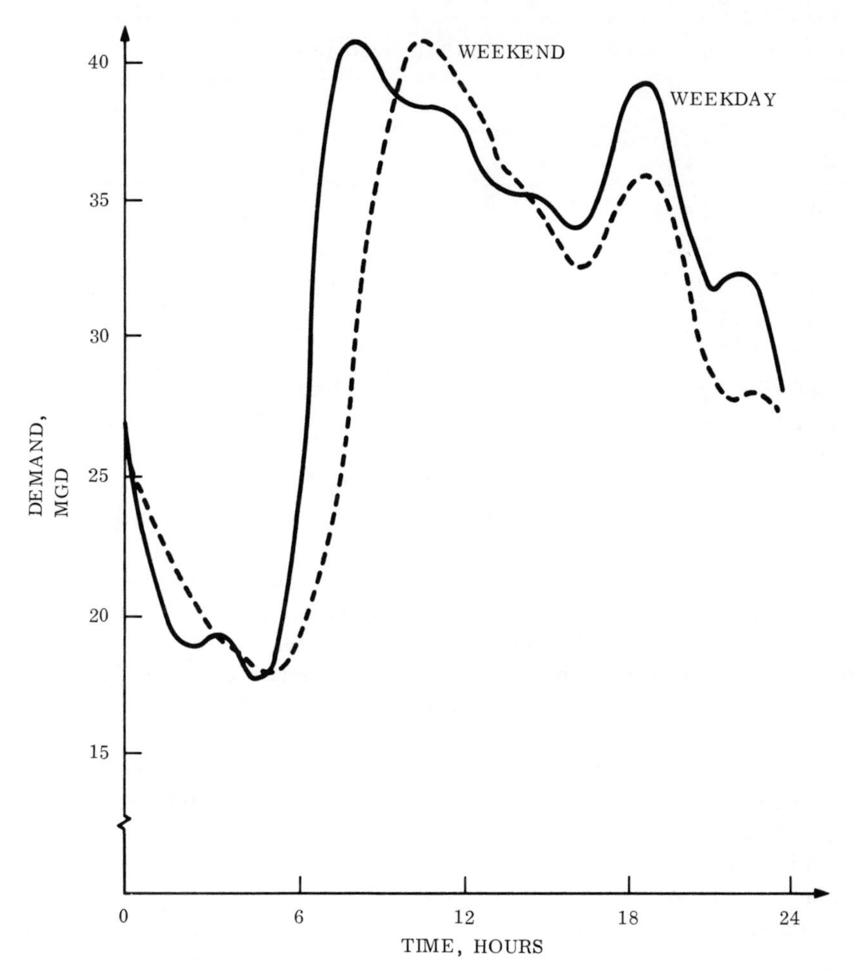

Figure 4-1. Average Weekday and Weekend Demand Curves

clear-weather loads are due to lawn sprinkling and, especially in the cities, to fire hydrants opened by children. For the present purpose, eight weeks of summer data were studied, broken up into weekday-weekend and rain-clear categories.

An effective way to include the effect of temperature is to modify the mean value function in Equation (4.2) as follows:

$$\widehat{Qd_s}(t;\, NH) = \left[\; a2_s(0) + \sum_{j=1}^{NH} a2_s(j)\,\cos(2\pi jt) + \sum_{j=1}^{NH} b2_s(j)\,\sin(2\pi jt) \right] TM^{PS}\,, \tag{4.30}$$

Table 4-1
Effect of Temperature on Residual Standard Deviation

		No temperature effect		*Temperature effect with optimum PS*	
		PS	$\hat{\sigma}_r$	*PS*	$\hat{\sigma}_r$
Weekday	Clear	0	4.59	1.38	3.74
	Rain	0	4.59	0.88	3.83
Weekend	Clear	0	4.08	0.80	3.94
	Rain	0	4.35	1.2	3.64

where TM represents maximum temperature and PS represents an experimentally found exponent. The subscript s refers to the summer estimation equation. In place of the periodic demand values $q(n)$ used to find the harmonic coefficient estimates in Equation (4.2), the values $[q(n)/TM^{PS}]$ are used in Equations (4.3) through (4.5) to obtain the estimates for Equation (4.30).

The optimum exponent PS can be found by a one-dimensional search. If we take a category, say clear weekdays, the average harmonic coefficients are found for all days within the category; then, with Equation (4.30) acting as the mean value function, overall residual variance is obtained. If $PS = 0$, daily maximum temperature would have no effect, while if PS is large, Equation (4.30) would be too sensitive to temperature. Therefore, intuitively, the optimum PS for a particular set of data lies somewhere in between. Table 4-1 shows the reduction of residual standard deviation caused by considering temperature.

Figures 4-2 and 4-3 show average summer clear and rain demand curves at 80°F maximum temperature for weekdays and weekends, respectively. Rain and clear curves are substantially the same except for the late afternoon and early evening. Presumably the reason for this is the curtailment of outdoor water use activities due to thunder storms.

Table 4-1 shows, for the optimum exponent PS, residual standard deviations, $\hat{\sigma}_r$, which are all less than 4.0, as compared to $\hat{\sigma}_r = 4.2$ for the winter data. This does not imply that summer demand can be predicted more accurately than winter demand, because Table 4-1 assumes perfect knowledge of temperature. To account for temperature uncertainty, the average daily consumption variance expression [Equation (4.29)] will be modified by the addition of a term. An exact analysis is impossible, so what follows contains approximations.

Average demand, Qd, can be divided into terms that relate to temperature and those independent of temperature:

$$\overline{Qd} = (TB + TR)^{PS} (QB + QR), \tag{4.31}$$

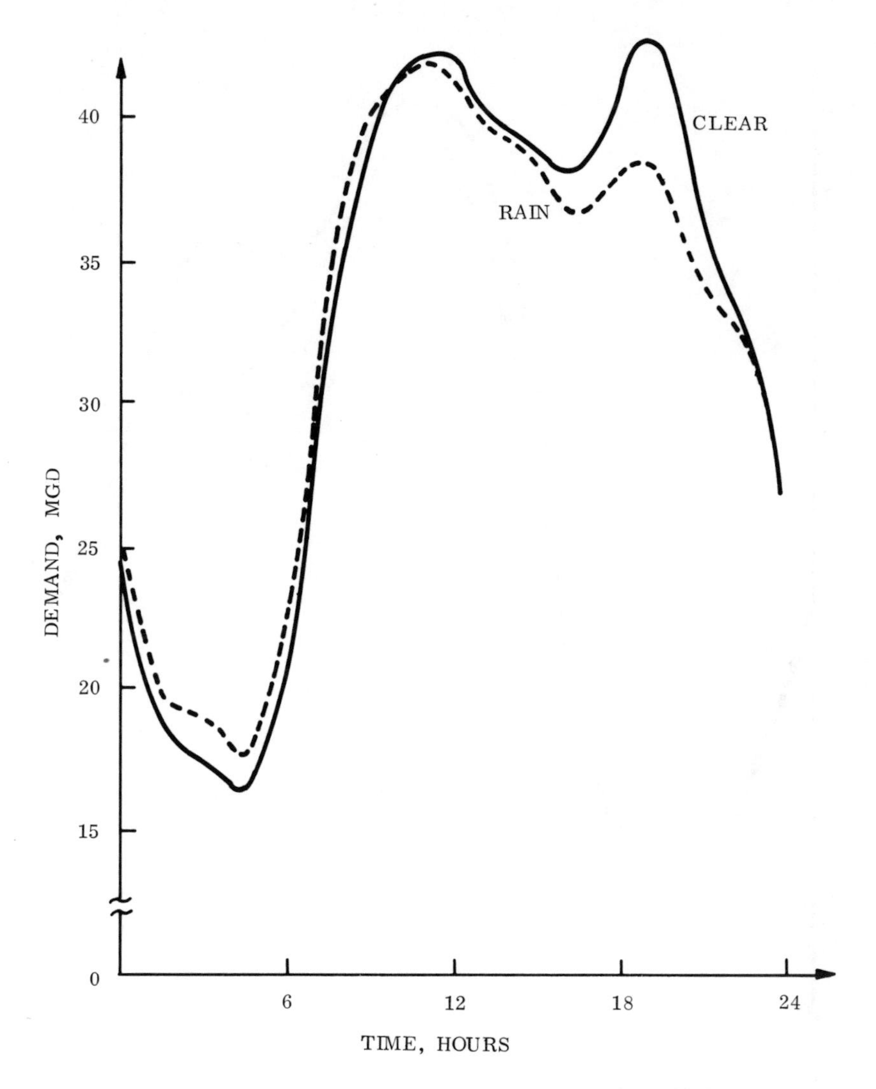

Figure 4-2. Summer Weekday Demand at 80°F

where TB represents the expected next day temperature; TR, the prediction error random component, variance σ_T^2; QB, the next day expected demand/temperature ratio, $QB = Qo/TB$; QR, the random component of QB, variance σ_{QB}^2. In the cases encountered, the exponent PS is nearly 1, so

$$\overline{Qd} \cong TB \cdot QB + TB \cdot QR + QB \cdot TR + TR \cdot QR. \qquad (4.32)$$

Then

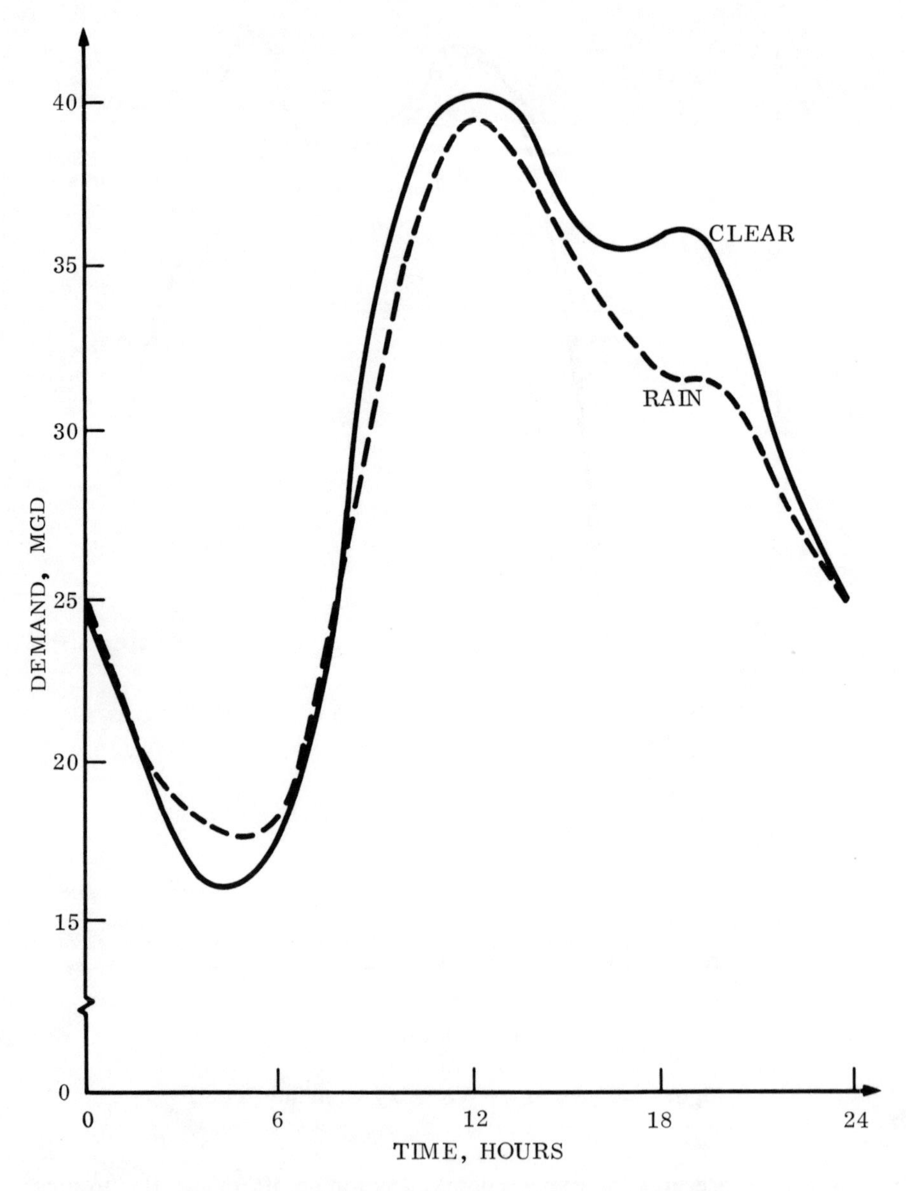

Figure 4-3. Summer Weekend Demand at 80°F

$$\mathrm{Var}(\overline{Qd}) \cong 0 + TB^2\sigma_{QB}^2 + QB^2\sigma_T^2 + \sigma_T^2 + \sigma_T^2\sigma_{QB}^2\,\mathrm{Var}(\chi_1^2); \qquad (4.33)$$

where χ_1^2 is the chi-square random variable (with one degree of freedom) whose variance is equal to 2.

$$\mathrm{Var}(\overline{Qd}) = \sigma^2_{Qd} \cong TB^2\sigma^2_{QB} + QB^2\sigma^2_T + 2\sigma^2_T\sigma^2_{QB}\,. \qquad (4.34)$$

If the maximum temperature is known exactly, σ^2_{Qd} is known from Equation (4.29). If we set $\sigma^2_T = 0$, it is possible to solve for σ^2_{QB}:

$$\sigma^2_{QB} = \frac{\sigma^2_r}{TB^2 \cdot NT}\,. \qquad (4.35)$$

If this is substituted back into Equation (4.34), we obtain

$$\sigma^2_{Qd} \cong QB^2\sigma^2_T + \frac{\sigma^2_r}{NT}\,(1 + 2\sigma^2_T/TB^2)\,. \qquad (4.36)$$

Consider now some nominal values: $\sigma^2_T = 1°\mathrm{F}$; $TB = 80°\mathrm{F}$; $Qo = 30$ million gallons per day. The second term in the parenthesis of Equation (4.36) is small compared to the first, so, with the alternate definition of QB:

$$\hat{\sigma}^2_{Qd} \cong \hat{\sigma}^2_r/NT + [(Qo/TB)\sigma_T]^2\,. \qquad (4.37)$$

This is the result sought: a modification of Equation (4.29) to account for temperature uncertainty. Through the use of the above nominal values with the particular data analyzed, the winter and summer total consumption standard deviations are as follows:

Winter: $\hat{\sigma}_{Qd} = 0.61$ million gallons;

Summer: $\hat{\sigma}_{Qd} = 0.74$ million gallons.

An idea of total consumption prediction accuracy can be obtained by remembering that the 95 percent confidence interval is 1.86 standard deviations.

These last results are not needed for the present control study and are presented as a matter of interest. One could arrive at the average daily consumption standard deviation without calculating a time varying mean value function. The important results of this chapter are the weekday and weekend total demand mean value functions.

5 Distribution Control

The objective of automatic control of water distribution is to minimize the pumping cost required to provide water to customers at adequate pressure. Figure 1-1 showed that in a tank-dominated district, customer pressure is largely determined by tank depth and customer elevation. For this reason, the problem of maintaining adequate minimum pressure is reduced to that of keeping tank depths above minimums that are assumed to be given for the present study. On the other hand, upper tank depth constraints are usually determined by physical tank dimensions.

Control variables are the discrete pump combinations that may be put on line at each pumping station, and the return function to be minimized is total pump energy consumption. There are special problems caused by effecting control by changing discrete pump combinations. In many cases, only a few combinations are available, so, at a time when it might be desirable to hold tank depths constant, one combination will pump too much, and the next too little, thus leading to limit cycling. For this reason, there must be an additional constraint to keep pump cycling at an acceptable level. This is rather a nebulous requirement. Some water distribution experts [7] prefer to make as few changes as possible, while others [33, 69] are willing to make several changes per pump per day. For the present purpose, the control will be considered satisfactory if the average number of pump changes is on the same order as current practice.

The method of control developed in this chapter is based on dynamic programming (DP). DP analysis will lead to time varying thresholds in tank depth, which, when crossed, cause pump changes to take place. *Dead bands* placed about the switching lines reduce pump cycling to an acceptable level. At the same time, dead bands render the control system relatively insensitive to the effects of random demand variations evident in the demand study in Chapter 4.

An ideal district is studied where all pumps are of the same specific speed (all pump head-flow curves have the same shape but are scaled to pump design flow) and identical efficiency curves. Step responses clearly show how DP causes an increase in water storage in anticipation of a demand increase, and how this effect is varied by changing pump efficiency curves. Finally, automatic control of an actual system is simulated over the period of one week. Energy consumption is compared to that required by a "floating tank" policy and to that consumed in the reference week, when the actual system was controlled by supervisory control.

Applicability of Dynamic Programming

While it has been shown possible to determine an expected demand function, random variations prevent the accurate prediction of the tank depths states with time. For this reason, an open-loop control is not acceptable [8, 48]. What is needed is a closed-loop control that will optimally transform the system from any initial state into the future, possibly to a specified end state. DP performs this function, guided by Bellman's principal of optimality: "An optimal policy has the property that whatever the initial state and initial decisions are, the remaining decisions must constitute an optimal policy with regard to the state resulting from the first decision" [3, 1972]. DP has several properties that make it especially suitable for application to the current problem. While the DP principles can be written as continuous equations, for practical purposes, they must be solved numerically using discrete time increments. The macroscopic model time simulation is already carried out in this way. Also, the current problem has time varying constraints which assure that a decision made at one point in time will not lead to a tank overflow or underflow at some time in the future. Rather than complicate the problem, constraints actually reduce the calculation required of DP.

As a brief review, the discrete deterministic DP equations are summarized. The notation is that of R.E. Larson [42, 1967], who presents a concise summary of various aspects of DP. System dynamics are described by the following vector nonlinear difference equation:

$$\mathbf{x}(n + 1) = \boldsymbol{\phi}[\mathbf{x}(n), \mathbf{u}(n), n], \tag{5.1}$$

where $\mathbf{x}(n)$ represents the system state at time stage n; $\mathbf{u}(n)$, the control vector; and $\boldsymbol{\phi}$, the vector function $[\mathbf{x}(n), \mathbf{u}(n), n]$. The return function over NT stages, J, is the sum of the individual return functions, L, at each stage in time:

$$J = \sum_{n=0}^{NT} L[\mathbf{x}(n), \mathbf{u}(n), n]. \tag{5.2}$$

States and controls are constrained:

$$\mathbf{x} \in \mathbf{X}(n), \tag{5.3a}$$

$$\mathbf{u} \in \mathbf{U}(\mathbf{x}, n), \tag{5.3b}$$

where $\mathbf{X}(n)$ is the set of admissible state at stage n, and $\mathbf{U}(\mathbf{x}, n)$ is the set of admissible controls at state $\mathbf{x}$, stage n. Given an initial state, $\mathbf{x}(0)$, the control sequence $[\mathbf{u}(0), \mathbf{u}(1), \ldots, \mathbf{u}(NT)]$ must be found such that J is minimized.

If the minimum cost function starting at stage n in state $\mathbf{x} \in \mathbf{X}$ and ending at stage NT is defined as

$$I(\mathbf{x}, N) = \min_{\substack{\mathbf{u}(j) \\ j=n, \ldots NT}} \left\{ \sum_{j=n}^{NT} L[\mathbf{x}(j), \mathbf{u}(j), j] \right\}, \tag{5.4}$$

then the mathematical equivalent to Bellman's principal of optimality is as follows:

$$I(\mathbf{x}, n) = \min_{\mathbf{u}} \{L(\mathbf{x}, \mathbf{u}, n) + I[\phi(\mathbf{x}, \mathbf{u}, n), n + 1]\}. \tag{5.5}$$

This states that the minimum cost for state $\mathbf{x}$ at stage n is found by choosing the control that minimizes the sum of the cost to be paid at the present stage plus the minimum cost of going to the end from the state at $n + 1$ that results from this control.

For computational purposes, the states as well as time must be divided into discrete intervals. The DP as described in Equation (5.5) requires calculation backward in time. At any stage n, the minimum costs $I(\mathbf{x}, n)$ and controls $\mathbf{u}$ are calculated only for the discrete states. Next, at stage $n - 1$, from each discrete stage and for each control $\mathbf{u} \in \mathbf{U}(\mathbf{x}, n)$, a projection is made forward in time to stage n. Generally, the state $\boldsymbol{\phi}(\mathbf{x}^*, \mathbf{u}^*, n - 1)$ resulting from the control $\mathbf{u}^*$ applied at state $\mathbf{x}^*$ will not land on one of the quantized states at stage n. For this reason, some form of interpolation is required to find the cost in transforming $\boldsymbol{\phi}(\mathbf{x}^*, \mathbf{u}^*, n - 1)$ to the end state in terms of the values $I(\mathbf{x}, n)$ at the neighboring quantized states. For any further details on DP, it is recommended that any of the DP references [3, 42, 65] be consulted.

To summarize, the DP calculation works backward in time. When completed, at each time stage and each quantized state, the optimum control will be specified. At this point, it is not possible to recognize an optimum trajectory from beginning to end. For this purpose, a reconstruction is necessary, consisting of a time simulation starting at the initial state $\mathbf{x}(0)$. If a continuous control were being used, the control would have to be interpolated at each time stage. For discrete controls, at each stage, the optimal control at the nearest state is used.

We shall now return to the distribution control problem. At each stage in time, the pumps to be on line at the pumping stations are specified at the quantized tank-depth variables. The water distribution problem is non-stationary because of the time varying demand forcing function and stochastic because of the component of demand that is random. In the theory of stochastic DP, the expected return function is minimized. The tank depth state $\phi(\mathbf{x}, \mathbf{u}, n - 1)$ is not a linear function of the state $\mathbf{x}$ at the previous stage $n - 1$, but, for short time intervals, for all practical purposes it is linear. While the expected value of the next state using a stochastic forcing function is generally not equal to the next state using the demand expected value, for half-hour integration intervals, they are essentially the same. Therefore, in the work to follow, deterministic DP will be used with the demand mean value function acting as the "known" future demand.

The most straightforward application of DP to water distribution control is to assign each tank depth to a state variable. For more than one or two tanks, this soon gets out of hand. DP is essentially a refinement of total enumeration; and, while it requires drastically fewer trials than total enumeration, the number of trials, associated computer storage, and computation time grow rapidly with additional state variables. Other difficulties include the need for multidimensional interpolation and increased complexity of constraints. There have been made various improvements on the basic DP procedure to reduce such problems, such as a memory requirement reduction by state increment dynamic programming. However, as Larson [42, 1967] points out, the successful application of DP often depends upon a clever problem formulation. It is for this reason that the approximation of a single state variable is presented in the following section.

Single State Variable Control of a Multiple Tank District

In a district where demand is substantially proportional, if pumps of the same relative strength are on line at each station, all tank depths tend to rise and fall together. This observation can be used to advantage since it permits analysis using a single state variable: equivalent tank head. The equivalent tank head is that head in each tank that would be reached if all pumping and demand were to cease, and tank head were allowed to equalize.

In going from multiple state variables to one, great savings in computer time and storage are attained, but obviously something is lost. If each tank were a state variable, DP would tend to draw more water from those pumps having higher than average efficiency. If one station were particularly efficient, its increased output would cause the head of the nearest tank to remain on the average above the head of a tank near a pumping station where less water was pumped. This effect is lost in going to one state variable if, in dealing with equivalent tank head, all tank heads are to be kept about the same. For this reason, it is necessary to introduce the concept of a *tank bias*, which is the average difference between each tank head and the equivalent tank head. Now, rather than have DP take care of the entire control problem, some of the responsibility is transferred to the analyst. Prior to implementation of DP control, DP simulations must be made with various tank biases in search of a minimum energy condition. In other words, a preliminary parameter search is needed.

For the purpose of holding tank heads to within a given bias of one another, pumps operating at all stations must be of the same relative strength. This means that when a pump change is needed, pump combinations will be changed at all pumping stations simultaneously so the resulting change in all tank depth rates will remain about the same. A balanced pump combination will refer to those pump combinations at all stations of the same relative strength. In the backward DP calculation, only balanced pump combinations are used. In the reconstruction, or in actual practice, balanced pump combinations will also be used so long as tank heads remain close to within the specified bias of the equivalent tank head. Just how close the heads must remain to the equivalent tank head will be referred to as the *tolerance.* If a tank falls below the allowed tolerance, a larger pump at the nearest station will go on to raise that tank depth. Similarly, if one tank head becomes too great, a local pump will cut back.

In the previous discussion, three parameters to be determined by the analyst have been described. They are summarized as follows:

1. *Dead Band:* A dead band surrounds each time varying switching line. The switching line divides equivalent tank depth into regions that require the operation of different balanced pumps. Pump changes then occur at the dead band extremeties, causing the control to become insensitive to short-term variations. The dead band must be determined by simulation such that, on the one hand, pump cycling is not excessive, and on the other, such that tank depth range and energy consumption are not excessive.

2. *Tank Head Bias*: In order to recover what was lost in going from many state variables to one, the bias—or average difference in tank head relative to one equivalent tank head—must be found. The correct bias will hold the head of the tank near the most efficient pumping station higher than the rest, thus causing more water to be pumped by the efficient station. Owing to the limited number of discrete pump combinations available at each station, there may be a very limited set of biases attainable.

3. *Tolerance:* Local pump changes will be required to bring tank heads within a tolerance of the specified biases. If the tolerance is too tight, many pump changes will be required; if, however, it is too loose, the tank heads will drift far from those used in the backward DP calculation, and the energy consumption will suffer.

A rigorous determination of these parameters would require a formidable multidimensional search, however, with a little experience with time simulations, a feel can be obtained for the sensitivity of various response characteristics to variation of the parameters. In the particular system studied, variation of these parameters causes considerable differences in pump cycling and tank depth excursions, but it brings about little change in energy consumption. Thus, it would appear that an analyst familiar with local control policies could select parameters that retain some of the previously used operating characteristics. At the same time, the DP control determines the timing of pump changes, which results in a significant reduction of pump energy consumption.

Dynamic Programming and Reconstruction Computation

The DP computation is carried out essentially as outlined in Equations (5.1) through (5.5). In this section, these general equations will be related to the specific equations of water distribution. Step by step details of the DP and reconstruction will not be spelled out, but they can be inferred from the flow diagrams in Appendixes B and C.

DP is carried out with equivalent tank head as the single state variable. The equivalent tank characteristic is given as a function of the individual tank characteristics as defined in Equation (2.13):

$$Cde = \left[\sum_{j=1}^{J} Cdr(j)^{-1} \right]^{-1} \text{ feet/million gallons} . \tag{5.6}$$

Then, the equivalent tank head, Hte, is a function of the individual tank heads:

$$Hte = Cde \sum_{j=1}^{J} \frac{Ht(j)}{Cdr(j)} . \tag{5.7}$$

In the terminology of Equation (5.1), the system state at time n is the equivalent tank head:

$$x(n) = Hte(n). \tag{5.8}$$

The control vector $\mathbf{u}(n)$ specifies which pump combination is on line at each pumping station. The state difference equation is then

$$Hte(n) = \phi[Hte(n - 1), \mathbf{u}(n - 1), n - 1], \tag{5.9}$$

where

$$\phi = Hte(n - 1) + Cte \sum_{j=1}^{J} Qt(j)\, \Delta t . \tag{5.10}$$

The tank flows, Qt [From Equation (2.8)], are indirect functions of the pumps on line, $\mathbf{u}(n - 1)$ or $k(i)$, and the time $n - 1$ through the demand mean value function [Equation (4.2)] for $t = (n - 1)/NT$.

The return function is the sum of energy consumed by all pumping stations:

$$L[Hte(n), \mathbf{u}(n), n] = Cpe \sum_{i=1}^{I} \frac{Qp(i) Hpu(i)\, \Delta t}{Fpe(i, k)} . \tag{5.11}$$

This expression follows directly from Equation (2.35), where Δt represents the integration interval $1/NT$: Qp, the pump flow from Equation (2.24); Hpu, the pump head from Equation (2.36); and Fpe, the efficiency from Equation (2.37). The control vector, $\mathbf{u}(n)$, consists of the pumps on line $k(i)$

that affect Fpe, Hpu, and Qp. The latter two are also affected by the time n through the demand function.

The set of admissible state variables $\mathbf{X}(n)$ does not depend on time n. There is an absolute upper limit ($Hteu$) and lower limit ($Htel$) that the equivalent tank head Hte must not cross. These limits are set from given tank depth limits required to maintain adequate pressure. Owing to the dead band and tolerance, individual tank heads, $Ht(j)$, can slightly exceed these limits.

The set of admissible controls $\mathbf{U}(\mathbf{x}, n)$ is a function of both the state $\mathbf{x} = Hte$ and the time $t = n/NT$. An upper time varying bound $Bu(n)$ in equivalent tank head must be defined above which only the smallest balanced pump combination may be used such that, at a future time, Hte will not exceed $Hteu$. Similarly, there is a lower bound $Bl(n)$ below which only the largest balanced pump combination may be used such that Hte will not fall below $Htel$ at a later time. Figure 5-1 illustrates the fixed state limits $Hteu$ and $Htel$, and the variable state bound $Bu(n)$ and $Bl(n)$. The regions in state Hte and time n are as follows:

I. Only largest balanced pumps may be used so that $Htel$ will not be violated.
II. Any balanced pumps may be used.
III. Only smallest balanced pumps may be used so that $Hteu$ will not be violated.

In order to find the time varying bounds, a backward time integration is necessary. If at time n, Hte is at the upper limit $Hteu$, then, when integrating backward with the smallest balanced pumps, the head must rise in order to keep the upper bound equal to the upper limit. If the head falls in backward integration, it will rise in forward integration, and thus the upper bound will be violated. For this reason, a backward difference equation must be defined similar to the forward equivalent, Equations (5.9) and (5.10):

$$Hte\,(n-1) = \phi'[Hte(n), \mathbf{u}(n), n], \tag{5.12}$$

where

$$\phi' = Hte(n) - Cte \sum_{j=1}^{J} Qt(j)\,\Delta t\,, \tag{5.12$'$}$$

The logic for finding the upper boundary is as follows: Using the smallest balanced pumps, find

$$Hte(n-1) = \phi'[Bu(n), \mathbf{u}, n]. \tag{5.13}$$

$$\text{If } Hte(n-1) \geqq Hteu, \text{ then } Bu(n-1) = Hteu. \tag{5.14a}$$

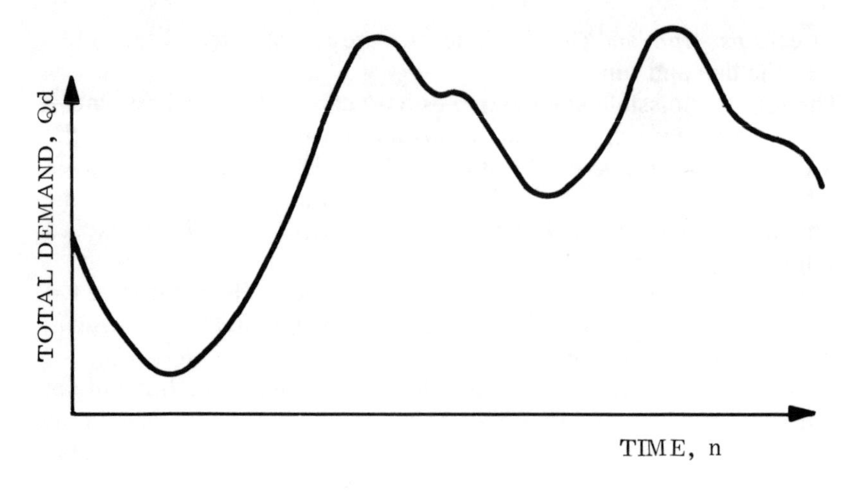

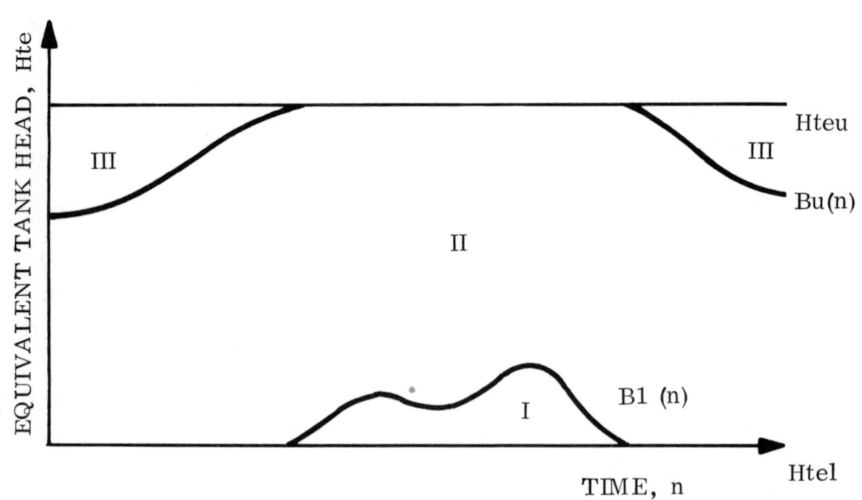

Figure 5-1. Equivalent Head Limits and Bounds Illustrated in Relation to Total Demand

$$\text{If } Hte(n-1) < Hteu, \text{ then } Bu(n-1) = Hte(n-1). \quad (5.14b)$$

Similarly, for the lower boundary, using the largest balanced pumps, find

$$Hte(n-1) = \phi'[Bl(n), \mathbf{u}, n]. \quad (5.15)$$

$$\text{If } Hte(n-1) \leqq Htel, \text{ then } Bl(n-1) = Htel. \quad (5.16a)$$

$$\text{If } Hte(n-1) > Htel, \text{ then } Bl(n-1) = Hte(n-1). \quad (5.16b)$$

Finally, there is the matter of interpolation. The return function $I(\mathbf{x}, n)$ in Equation (5.4), for computational purposes, can take on values only at discrete increments of the state variable $\mathbf{x}$, which is, in this case, the scalar $x = Hte(n)$. The return is energy consumption, and is, for the purpose of discussing interpolation, expressed as a function of the quantized state x_h.

$$E_h = f(x_h) \tag{5.17}$$

A straight-line interpolation from E_h to E_{h+1} proved unsatisfactory because of very slight variations in slope from one segment to the next. A least-squares fit of an exponential curve to the energy points E_h produces a continuously varying slope and eliminates the instability problem experienced using a straight-line interpolation. The general form of the exponential curve is

$$\hat{E}_h + Eo = C_1 \, e^{C_2(x_h - x_o)}, \tag{5.18}$$

or, equivalently,

$$\hat{E}_h + Eo = Ca \, e^{Cb \, x_h}. \tag{5.19}$$

The three parameters Eo, Ca, and Cb must be determined from the values E_h and x_h. This expression is not linear in the three constants, so the constant Eo may be found by trial and error, a process that may be done formally by a dichotomous search [84]. At each stage in the search, for an assumed Eo, the values Ca and Cb are found by the regression normal equations; then the sum-square error,

$$\sum (E_h - \hat{E}_h)^2,$$

is found. The optimum Eo minimizes this sum square. The resulting exponential equation can approximate a curve containing almost a right angle to nearly a straight line of any slope. An example of an exponential data fit is shown in Table 5-1.

Customarily, DP serves the purpose of providing an optimal path in state space from an arbitrary beginning state to a specified end state. For the current problem, this would require the specification of the equivalent tank head at sometime in the future, but what would be the basis for selecting this head? Fortunately, the backward DP calculation reaches a steady state rapidly; therefore, if, for example, the backward DP calculation is made for three days that have the same expected demand, after there is an initial transient, the first two days' results are the same. For this reason, rather than arbitrarily specifying an end state twenty-four hours hence, the DP in steady state will smoothly pass from one day to the next with no specific end states specified. For further details of the backward DP

Table 5-1
Energy Values (in 1000 Kilowatt-Hours) for Exponential Curve Approximation of Quantized State Energy

Actual	6.84	6.06	5.32	4.62	3.94	3.30	2.69	2.10	1.54	1.00	0.49	0.00
Exponential	6.78	6.04	5.33	4.64	3.98	3.34	2.73	2.13	1.56	1.00	0.47	−0.05

calculation, see Appendix B, which shows the computer program flow diagram.

In the discussion that follows, refer to Figure 5-2. The result of the backward DP calculation is the division of the equivalent tank head-time space into regions where each balanced pump combination is to be used. In Figure 5-2, there are two "switching lines"; these indicate that three balanced pump combinations are available. The largest balanced pump combinations are to be used below the lower switching line, and the smallest are to be used above the upper line. The figure is an actual simulation of an idealized system to be described in the next section, so, to be specific, Table 5-2 shows the three balanced pump combinations.

In Figure 5-2, a two-foot dead band extends above and below each switching line, so a change in balanced pumps takes place only after the equivalent head extends two feet beyond a switching line. Pump changes [2] and [3] take place for this reason.

If a particular tank exceeds the tolerance from the equivalent tank head, a pump change must be made at the station that will most affect the offending tank. If tank $\tilde{j}$ is out of tolerance, a pump change is made at station $\tilde{i}$ corresponding to

$$\max_{i} \frac{\partial Qt(\tilde{j})}{\partial Qp(i)} = \max_{i} Cqt[\tilde{j}, (2 + i + j)], \tag{5.20}$$

where Cqt is given in Equation (2.8). In the specific example, pump station 1 most affects tank 1; similarly, pump station 2 most affects tank 2. This is not unreasonable considering their physical proximity, as shown in Figure 2-5. Now, with respect to Figure 5-2, the pump combinations on line after pump change [1] are—from Table 5-2—balanced combinations, but those on line before pump change [1] are not. The tolerance in this case was set at two feet. It can be seen that tank head 2 is more than two feet below the equivalent tank head, so the pump combination on line at station 2 is combination 2, one combination above what would be called for given the relative position of the equivalent head and the upper switching line. Pump change [1] takes place when tank head 2 momentarily crosses the equivalent head. Appendix C contains the flow diagram containing pump selection logic.

Table 5-2
Station Pump Combinations Which Combine to Form Balanced Pump Combination

Pump Station Number	*Balanced Pump Combination*		
	No. 1, smallest	*No. 2*	*No. 3, largest*
No. 1	No. 1	No. 2	No. 4
No. 2	No. 1	No. 2	No. 3

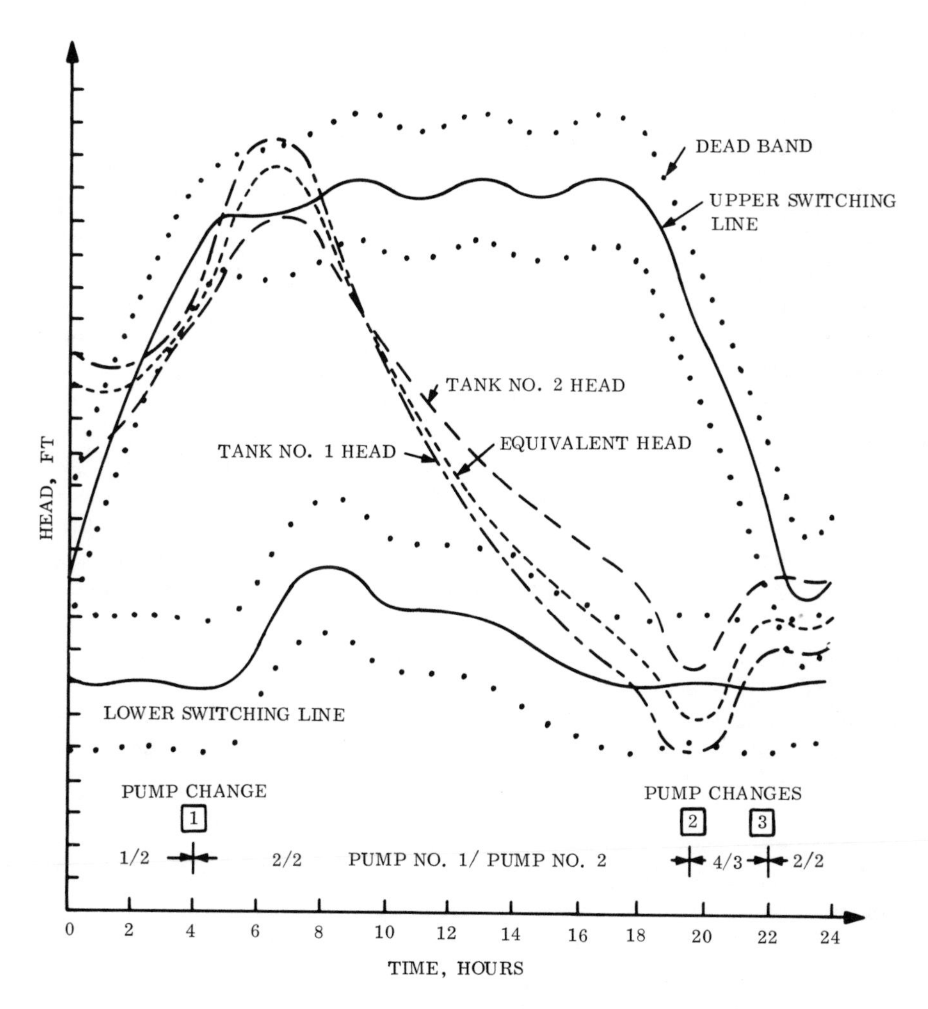

Figure 5-2. Dynamic Programming Control Simulation

Table 5-3
Torresdale-Fox Chase Design Flow and Average Flow (in million gallons per day)

Station	*Pump combination*	*Design flow*	*Average flow*
Station 1	1	10	12
	2	16	18
	3	20	24
	4	23	27
Station 2	1	6.3	7
	2	9.5	10
	3	15.8	14

Dynamic Programming Control Characteristics

The successful reduction in pump energy consumption by use of DP control results from the determination of the timing and quantity of water forced into elevated storage. These effects are more easily demonstrated in a somewhat idealized system, as will be shown in this section. The Torresdale-Fox Chase district will be dealt with as it presently exists in the next section.

Primarily as a result of the expansion of the Torresdale-Fox Chase district, most of the pumps are not matched to the system. Specifically, most pumps are operating in a high flow mode, far beyond their design points. Table 5-3 shows the design flow, Qpd for the various pump combinations, as well as the current average flow. For the present purpose, the pumps will be artificially matched to the system by assuming that the design flow Qpd is equal to the average flow as shown in Table 5-3. The result will be that the pumps will spend about equal time operating before and beyond the design point as shown in Figure 2-2. All pumps will be assumed to have the same specific speed, so the pump head-flow curves all have the same shape, scaled to the design flow Qpd. The pump head-flow equation [Equation (2.1)] is repeated, except here (Qp/Qpd) is raised to the exponent 1.85:

$$Hp/Hpd = C+(1-C)(Qp/Qpd)^{1.85} . \qquad (5.21)$$

With the design heads Hpd at the two pumping stations set at 370 feet and 135 feet, respectively, and with $C = 1.33$, all pump curves are defined, and the conversion to the format in Equation (2.3) is immediate. The standard pump head-flow curve appears in Figure 5-3.

Pump efficiency curves will also be standardized such that efficiency is a quadratic function of the ratio of pump flow to pump demand flow:

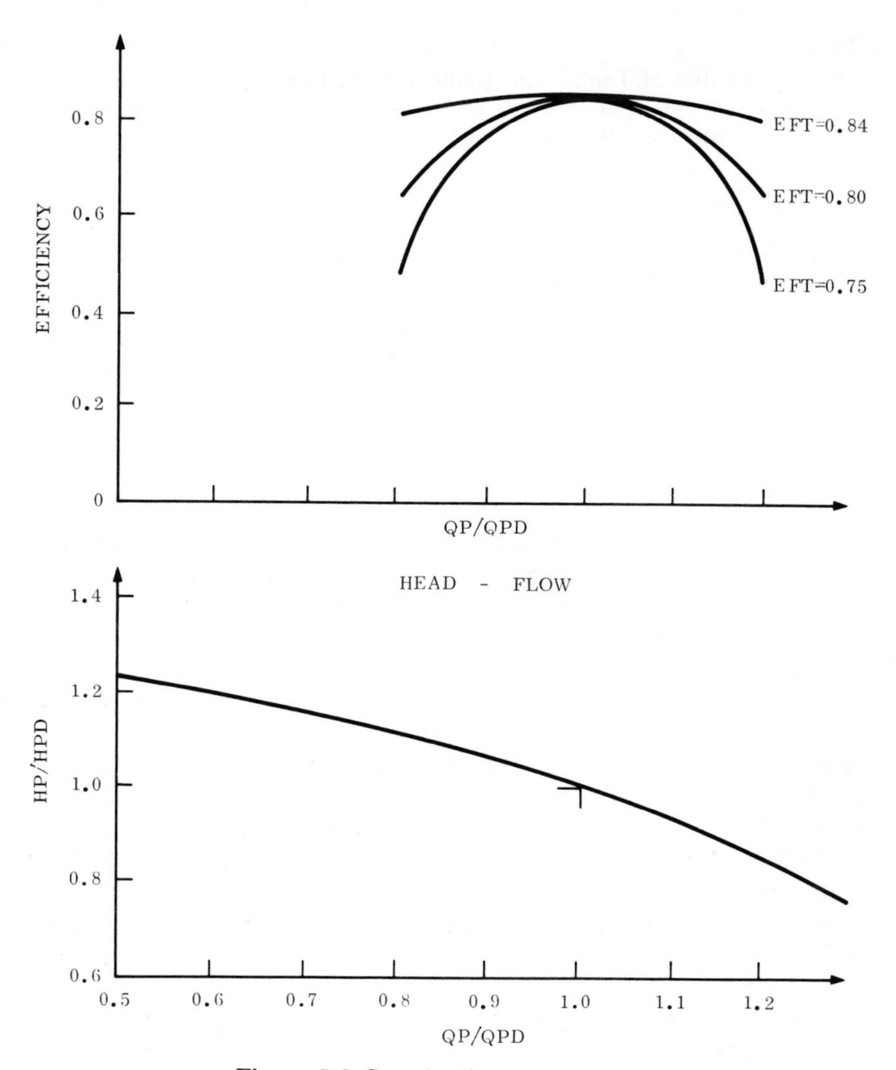

Figure 5-3 Standardized Pump Curves

$$Fpe = EFM - 100(EFM - EFT)[(Qp - Qpd)/Qpd]^2 . \qquad (5.22)$$

The efficiency function in Equation (5.22) is set up such that the maximum efficiency, *EFM*, is reached when the pump flow is equal to the pump design flow. Furthermore, the efficiency is equal to *EFT* when pump flow is 10 percent above or below the design flow. In Figure 5-3, the maximum efficiency is $EFM = 0.85$. Three curvatures are shown for $EFT = 0.84, 0.80$ and 0.75. It will be shown how DP causes the buildup of elevated storage as efficiency curvature is varied by changing the parameter *EFT*.

Table 5-4
Equilibrium Bias of Tank Head 1 Above Tank Head 2

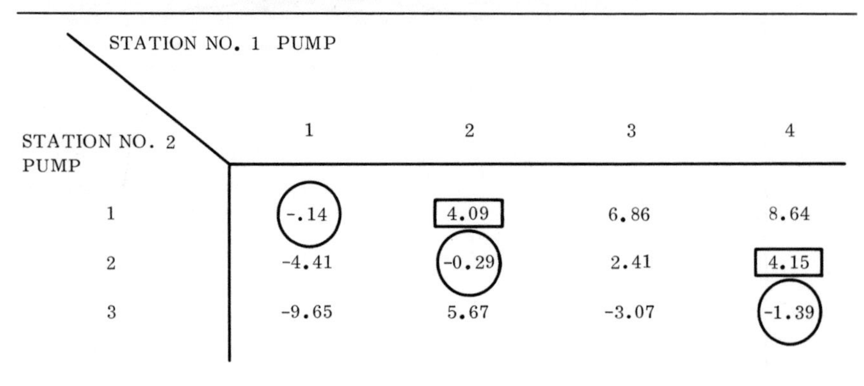

STATION NO. 2 PUMP \ STATION NO. 1 PUMP	1	2	3	4
1	-.14	4.09	6.86	8.64
2	-4.41	-0.29	2.41	4.15
3	-9.65	5.67	-3.07	-1.39

Balanced pump combinations can be chosen by calculating equilibrium tank depths and by noting differences in tank head as a function of pumps on line. For this specific example, Table 5-4 shows these head differences. Balanced pump combinations result in equilibrium tank heads, all of which differ by nearly the same bias. It could be possible to assign tank head 1 to remain about four feet above tank head 2, in which case the two balanced pump combinations would be (2, 1) and (4, 2), corresponding to those numbers in Table 5-4 enclosed by a box. For the present demonstration, a zero bias will be used, with the three balanced pump combinations being (1, 1), (2, 2) and (4, 3), corresponding to the encircled values in Table 5-4. These balanced pump combinations were previously shown in Table 5-2.

Figure 5-4 shows the switching line step responses of the idealized system as they are affected by the curvature of pump efficiency curves. The smallest balanced pumps are to be used above the upper switching line, while the largest are to be used below the lower switching line. The figure shows that the switching lines rise sooner and remain higher as efficiency curvature increases, or, equivalently, as *EFT* decreases. In a time simulation, this has the effect of building up the tank depths sooner and to a greater depth because, with increased curvature, a greater penalty is paid by operating the pumps far from their design points. At the higher demand, the lower the tank depth, the farther from the design points the pumps will operate.

The step response was shown in the form of switching lines rather than time simulations of tank depth. Most often, at a particular constant demand, one balanced pump combination will have an equilibrium tank head above the upper limit, while the next smaller will have an equilibrium head below the lower limit. Therefore, in order to maintain tank head within limits, the system will limit cycle with frequency and amplitude which is a

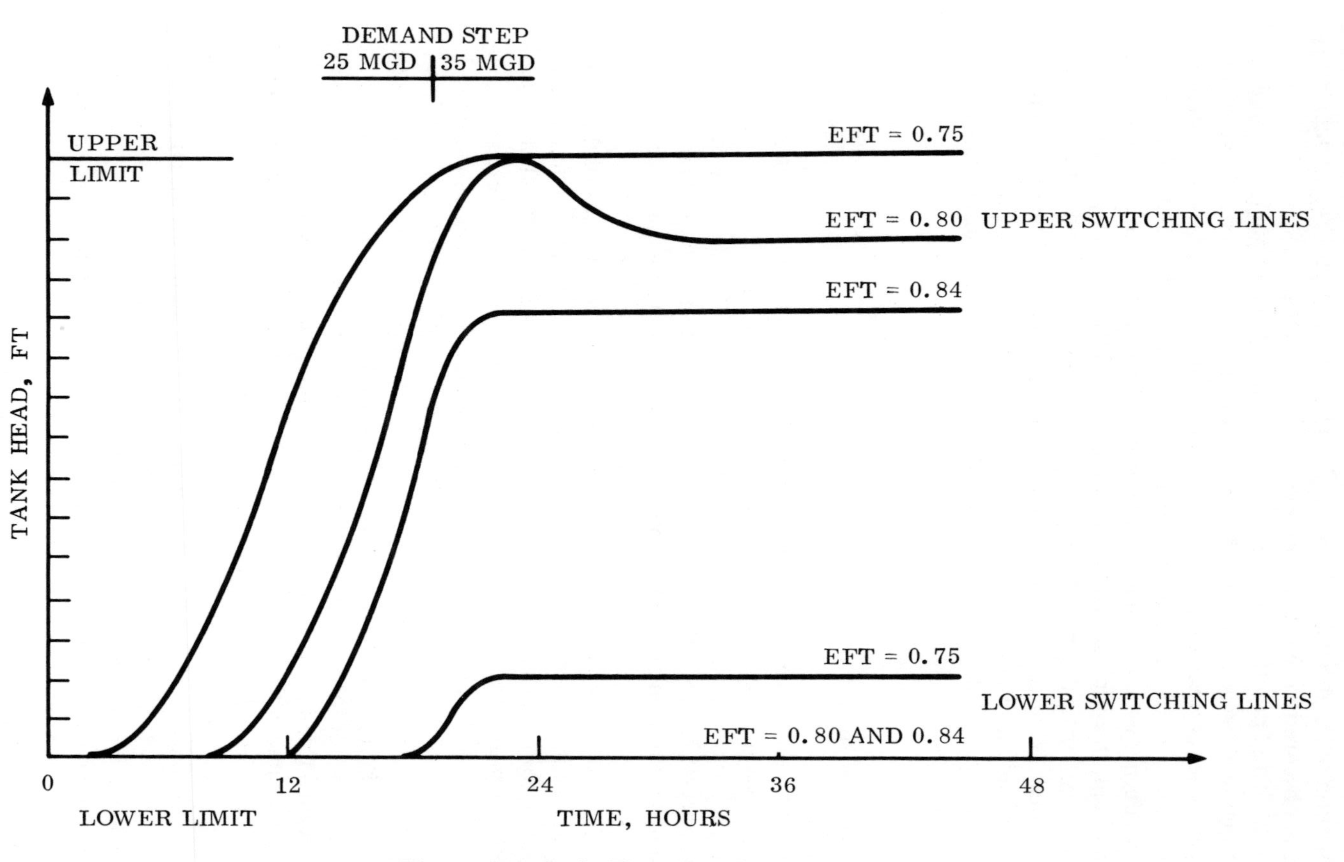

Figure 5-4. Switching Line Step Response

function of dead band and tolerance. The effect of the step in demand is thus obscured in the sawtooth cycling.

In actual practice, demand never remains constant but is continuously changing, as was shown in Chapter 4. For this reason, a steady-state limit cycle is never reached. Figure 5-5 shows a time simulation of tank 1 depth in relation to total demand. Although the simulation had been run several days during which no steady state was reached, the DP characteristics are apparent. In anticipation of a demand peak at 08:30, tank depth peaks a few hours earlier, as it does before the later demand peak at 19:00. As in the step response, the greater efficiency curvature (smaller *EFT*) leads to a greater depth buildup. In this simulation, dead band and tolerance were set at two feet. The tank 2 depth profile is not shown but is much the same as that for tank 1.

Control of Actual System

To attain the initial goal of controlling a pressure district without modification, a simulated control of the Torresdale-Fox Chase district was carried out. Demand as it actually occurred on the week of 10 January through 16 January 1972 was used as the forcing function. The important bases of comparison from one control method to another are the tank depth excursions, frequency of pump cycling, and total pump energy consumption.

In order to assist in evaluating the DP control, the results of the supervisory control are presented exactly as the system operators controlled the system for the reference week. A third means of control, the floating tank (FT), is presented. Under the FT policy, a policy that is presently advocated as an ideal by many, as few pump changes are made as possible. In this case, a basic set of pump combinations, whose total combined flow is approximately equal to the expected daily average consumption rate, is put on line. Pump changes are made only when tanks appear likely to violate their upper or lower limits. In the present case, both tanks have essentially a twenty-five-foot operating range, so pump changes were made if tank depths came within five feet of the upper or lower limits. A one-foot dead band prevented cycling about the switching point.

For the control simulations, the pump head-flow and efficiency curves were used as shown in Figures 2-6 and 2-7. For the backward DP calculation, only two demand expected value curves were used—the weekday and weekend curves developed in Chapter 4. A backward DP calculation through an entire week was not necessary because of the rapid attainment of a steady state. Although there were slight transitions in the switching lines from weekdays to weekend days and vice versa, for practical purposes, there were only two sets of switching lines, which corresponded to the two expected demand functions.

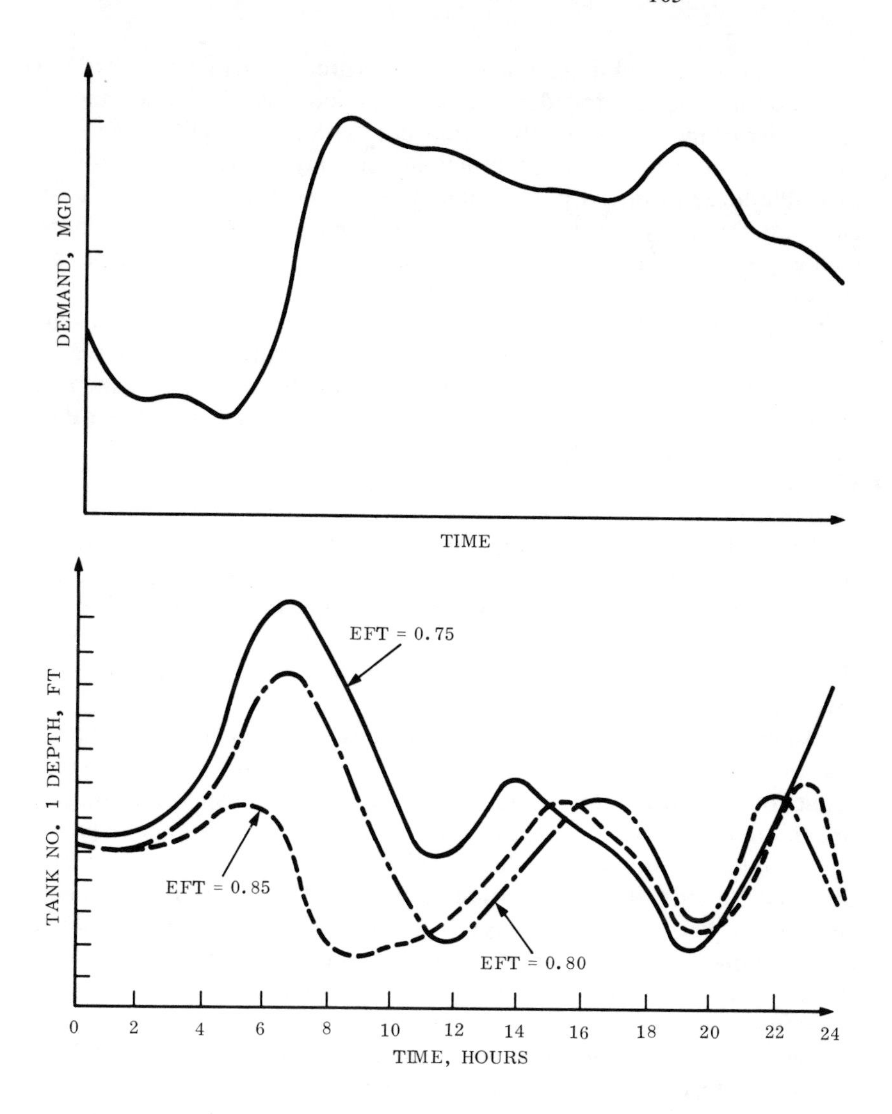

Figure 5-5. Tank Depth Time Response Related to Total Demand

Selection of tank head bias must be based on the determination of which are the relatively more efficient pumping stations, coupled with a knowledge of what biases are possible with the given set of pumps. In this case, a six-foot bias of the Fox Chase tank over the Somerton tank was chosen. The dead band and tolerance parameters tend to affect energy consumption little, while greatly affecting tank depth excursions and pump cycling. Suitable operating characteristics were obtained in this case using a three-foot dead band and tolerance.

The resulting week-long time simulations are shown in Appendix D. At the beginning of the first day, initial conditions are the same, and the remainder of the day is largely a transient, with FT and DP falling into a rather regular pattern for the remaining six days. The DP control had available three balanced pump combinations, but the middle one in this case was relatively inefficient and never chosen. For this reason, pump changes were made from relatively weak to relatively strong pumps, and the cycling is quite apparent from Tuesday on. Although it appears that there are three cycles per day, in fact there are only two. The first daily peak near 06:00 takes place when the weaker pumps are on line and the low early morning demand is in transition to the rapidly growing demand after 06:00. The later depth peaks near 15:00 and 21:00 result from the twice daily pump cycling. Note that the amplitude of the Somerton (tank 1) depth variations is greater than those at Fox Chase (tank 2). This is consistent with the slope II in the phase plane diagram (Figure 2-21).

Under DP control, the Fox Chase tank depth never falls below the five-foot level that was reached by the supervisory control on Tuesday. The Somerton tank depth remains generally below the supervisory control depths and comes close to the sixty-foot lower limit. There is no danger of the tank emptying since sixty-feet of water remain below this arbitrary lower limit.

Table 5-5 summarizes the number of daily pump combination changes and the total energy consumption. The DP control shows a 4.9 percent reduction in energy consumption relative to that consumed during the operation of the system for the reference week. The percentage reduction is, of course, a function of how well or poorly the system had been operated. Another factor to consider is how well the pumps are matched to the system. Part of the DP energy reduction results from holding a lower average tank head. This, in turn, causes the pumps to operate in a higher flow, lower head mode. If, as in the present case, the pumps were already operating at flows greater than their design flows, the reduction in energy caused by lower average head would be partially offset by a reduction in efficiency. In other words, had the pumps been better matched to the system, DP would have shown a greater reduction in energy consumption.

If the 4.9 percent saving were attainable throughout the year, at an average of $0.02 per kilowatt-hour, the annual saving would be over $10,000 in this district. Pump maintenance cost would not be affected by the increased number of pump changes if a policy of maintenance at regular intervals in time were in effect.

Another advantage, beyond the energy consumption reduction, is that the DP control provides customers with more constant water pressure, which results from the fact that tank depth excursions are smaller. Emergency reaction is also adequate since, if a main break were to occur,

Table 5-5
Control Characteristic Comparison

	Control		
	Dynamic programming	*Supervisory control*	*Floating tank*
Average daily pump changes:			
Torresdale	4	3	2
Fox Chase	4	2	0
Total week energy consumption (in kilowatt-hours)	209,918	220,661	222,689
Percentage change relative to supervisory control	−4.9	0	+0.9

the resultant tank level drop would cause larger pumps to go on line when a switching line was crossed.

The fact that DP does not make use of the upper fifteen feet of the tanks does not necessarily mean that this tank capacity is unnecessary. The present simulation was made in a period in January. Quite possibly the greater summer demands would require the use of the upper ranges of tank depth to prevent violation of tank lower limits.

The implication of the current work is that, while water departments may find some properties of a floating tank policy beneficial, in general such a policy will not lead to the lowest operating cost. Those publications that advocate floating tank do so when comparing it to a policy of holding tank depths at the upper limits for part of the day. Certainly, floating tank is cheaper than this policy, but intuitively it cannot be generally the cheapest. Pumps can be exactly matched to a system only at a specific demand, and for a set of specific tank depths. As conditions change during the day, it only stands to reason that it would be cheaper to make a few pump changes to accomodate current conditions rather than to try to hold one combination that accomodates day-long average conditions.

For those water distribution system operators not prepared to make such a detailed study as has been carried out here, savings may be attainable by holding tank depths more constant. A survey of pump station operating costs will point out which stations are less efficient; with this information, it will be clear that the average operating depths of the tanks near these stations should be lowered, depending of course on local constraints. In this way, it is possible for operators simply to attempt to gain from the DP control study, if they show the same willingness to experiment as those in the past who have tried to follow the floating tank policy.

6 Summary and Conclusion

Macroscopic Modeling

For the purpose of carrying out control studies, a macroscopic model has been developed that relates the major flows and pressures associated with pumps and tanks. The model consists partially of empirical expressions, including ones that calculate head drops from pumping stations to tanks, flows into tanks, and internal pressure points. The remainder of the model consists of conventional pump head-flow relations and tank depth-volume relations. The macroscopic model is made to produce a time simulation by relating a series of static solutions by numerical integration of flows into tanks to obtain tank depths. Total demand acts as the forcing function, and the pump combinations on line at each station act as control variables.

The model is validated by carrying out a time simulation using the same time varying demand and pump changes as took place during actual system operation. The model accuracy is determined by its ability to replicate actual operating data.

It may be concluded that, because of its speed and accuracy, the model is suitable for the purpose of control. Computation time is orders of magnitude less than that required by a corresponding network model primarily due to the fact that many times fewer variables are calculated and that a single static balance is noniterative. Accuracy results from the fact that the model parameters are found by regression analysis of actual operating data. With this method, no assumption need be made about unknown nodal consumptions or pipe roughness coefficients. For a two-pumping-station and two-tank system, the model has been shown to be essentially the exact equivalent of a network model producing a time simulation under proportional loading.

The fidelity of the model deteriorates somewhat as loading becomes less than proportional, but, for the largely residential pressure district studied, validation accuracy was still quite good. Further research will be required to show to what extent the macroscopic model will remain useful as it is applied to districts of increasingly industrial and, therefore, nonproportional loading. The model generality could be increased by the inclusion of in-line booster pumps and pressure reducing valves. Finally, in an effort to simplify the entire control problem, it may be possible to determine an approximately equivalent linear or piecewise linear model by examination of the phase plane characteristics.

Adaptive Modeling

The macroscopic model must be able to adapt continually to changes resulting from network modifications, to hydraulic changes due to pipe clogging or cleaning, and to changes in loading patterns. The Kalman filter has been adapted to carry out recursive regression. By taking adavantage of the "bandwidth" characteristic of the filter, one can provide a model at the end of each day that is suitable for use in simulating the next day.

It has been shown, for a week-long set of data, that the model, when adapted in this way, remains more accurate than if it were updated by daily batch regressions. Significantly, the model adapts without programming changes that would be required for a full network model. Less computer storage is required to carry out recursive as compared to batch regression.

The Kalman filter bandwidth must be periodically adjusted by analysis of a specific set of data. A useful area of study would be the investigation of means to adjust the bandwidth continuously.

Demand Study

The primary purpose in analyzing the time varying nature of total demand is to isolate predictable periodicities. This result, in the form of a mean-value function, is needed for dynamic programming control studies. A statistical test has been carried out that determines the number of harmonic coefficients sufficient to characterize the demand adequately. Another determines those mean value functions, corresponding to different days of the week, that are significantly different from one another.

For the data studied, it was found that six harmonics of one day are significant, and that weekday demand curves are significantly different than weekend demand curves. By including the effects of temperature and weather, we found that it was possible to reduce prediction error of summer data to the level of winter data error. Analysis of residual random variations showed the important result that the noise is uncorrelated at the one-half-hour interval. This result implies that a control based on frequent short-term projections into the future will be unsatisfactory because such a control will cause excessive pump cycling.

Further analysis of demand will be necessary to determine which seasons or parts of seasons of the year have significantly different demand characteristics. Furthermore, by considering the stochastic nature of demand, one could find tank depth minimums. There will be a point where the expected cost saving that results from lowering system heads is offset by the expected loss arising from the possibility of having to put large pumps on line, which would in turn incur a large penalty cost due to breaking the power demand limit.

Control

The goal of efficient water distribution control in a district containing elevated storage tanks is to minimize pumping cost while maintaining tank depths within set limits. Control is implemented by dynamic programming using a single state variable: equivalent tank head. Control variables include the discrete pump combinations on line at each pumping station. Constraints are tank head limits, and the return function to be minimized is pump energy consumption.

Working backward in time, using the demand mean value function as the forcing function, the time-state-variable space is divided into regions that indicate the most efficient balanced pump combinations to be used. The regions are separated by switching lines that are surrounded by a dead band in which no pump changes are made. The dead bands make the control relatively immune to random uncorrelated demand variations.

Simulations forward in time show the way in which a pressure district will react to dynamic programming control. Specifically, it has been shown that, as the curvature of pump efficiency curves increases, DP will cause a greater buildup in storage to avoid a later costly condition when pumps would operate in an inefficient high flow mode.

A week-long simulation has shown a 4.9 percent reduction in energy consumption when the DP control is used as compared to actual operation. DP control required about the same number of pump changes and held the system pressures more constant. Finally, it was shown that the often advocated floating tank policy is not generally the most efficient.

Appendix A: Regression

The purpose of regression is to relate one or more independent variables to a dependent variable. In linear regression

$$y_j = b_0 + \sum_{i=1}^{p} b_i x_{ji} + e_j, \qquad 1 \leq j \leq n, \tag{A.1}$$

where y_i is the value of the dependent variable at increment j; x_{ji}, the value of independent variable i at increment j; p, the number of independent variables; and e_j, the error term. In this equation, it is desired to find the regression coefficients b_i that minimize the sum-square error:

$$SSE = \sum_{j=1}^{n} e_j^2 . \tag{A.2}$$

The following matrices and vectors simplify the notation:

$$\mathbf{Y} = (y_1\ y_2\ \ldots\ y_n)^T, \tag{A.3}$$

$$\mathbf{B} = (b_0\ b_1\ \ldots\ b_p)^T, \tag{A.4}$$

$$\mathbf{e} = (e_1\ e_2\ \ldots\ e_n)^T, \tag{A.5}$$

$$\mathbf{X} = \begin{bmatrix} 1 & x_{11} & x_{12} & \ldots & x_{1p} \\ 1 & x_{21} & x_{22} & \ldots & x_{2p} \\ \vdots & \vdots & \vdots & & \vdots \\ 1 & x_{n1} & x_{n2} & \ldots & x_{np} \end{bmatrix} . \tag{A.6}$$

The matrix equivalents of Equations (A.1) and (A.2) are

$$\mathbf{Y} = \mathbf{XB} + \mathbf{e} \tag{A.7}$$

and

$$SSE = \mathbf{e}^T\mathbf{e} \tag{A.8}$$

It can be shown that Equation (A.8) is minimized if

$$\mathbf{B} = (\mathbf{X}^T\mathbf{X})^{-1}\mathbf{X}^T\mathbf{Y}. \tag{A.9}$$

As a practical matter, Equation (A.9) must be evaluated by a carefully designed computer program because of the fact that (X^TX) is frequently ill-conditioned, which leads to a poor numerical inversion.

The total sum-square deviation of the dependent variable is given by

$$SST = \sum_{j=1}^{n} (y_j - \bar{y})^2, \tag{A.10}$$

where

$$\bar{y} = \frac{1}{n}\sum_{j=1}^{n} y_j . \tag{A.11}$$

A single figure of merit of the ability of the independent variables to determine the value of the dependent variable is the multiple correlation coefficient:

$$\rho = \left(\frac{SST - SSE}{SST}\right)^{1/2} \tag{A.12}$$

As the regression improves, the sum-square error, SSE, diminishes and ρ approaches unity. If there is only one independent variable, the multiple correlation coefficient can be shown to be equivalent to the simple correlation between the two variables.

A statistical test is available to determine if the regression coefficients **B** are different from zero. The F ratio is formed:

$$F = \frac{(SST - SSE)/p}{SST/(n-p-1)} . \tag{A.13}$$

If the calculated F ratio is less than the tabulated value for the appropriate degrees of freedom(p) and $(n - p - 1)$ at a given level of significance, it cannot be said that the regression coefficients are different from zero. This is equivalent to saying that, at the given level of significance, it cannot be said that the relationship between dependent and independent variables is anything more than random. In a similar manner, a test can be made for each member of **B** to see if each is significantly nonzero.

Stepwise regression is merely a formal order in which statistical tests are applied in order to include only those independent variables that contribute significantly to the regression. If the level of confidence α is being used, only those regression coefficients are retained whose probability of being nonzero is $(1 - \alpha)$. Stepwise regression involves two tests:

1. *Enter-New-Variable test:* One at a time, each independent variable not currently a part of the regression is placed into the regression. For each, the probability is calculated that, were this variable part of the regression, its coefficient would be nonzero.

2. *Remove-Old-Variable test:* For all of those independent variables in the regression, the probability that each of their coefficients is nonzero is calculated.

Stepwise regression begins with no variables in the regression. Step by step, variables are added to the regression one at a time. The order of addition of these independent variables is determined by which had the greatest F ratio indicated by the enter-new-variable test, and they are added to the regression as long as their probability is greater than $(1 - \alpha)$. At each step, the remove-old-variable test is made to remove any independent variable whose probability has fallen below $(1 - \alpha)$. The procedure terminates when all remove-old-variable test probabilities are greater than $(1 - \alpha)$ and when all the enter-new-variable test probabilities are all less than $(1 - \alpha)$.

Appendix B: Optimization Flow Diagram

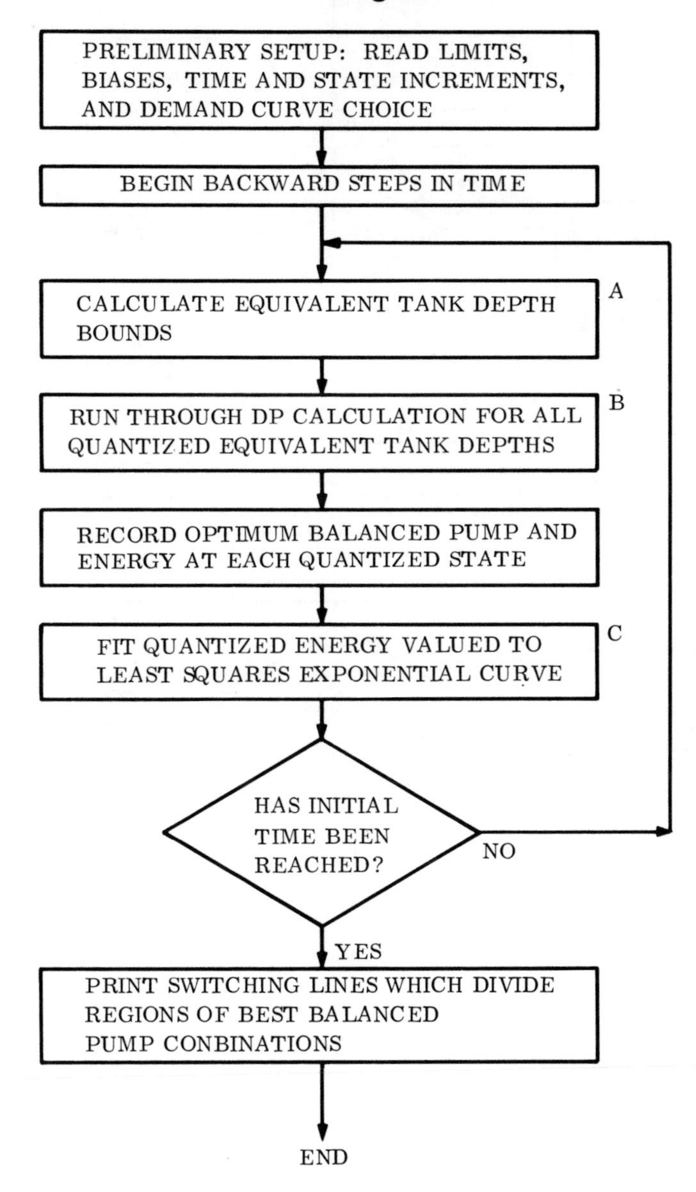

Details A, B, and C of this diagram are found on the following pages.

Figure B-1. Dynamic Programming Flow Diagram

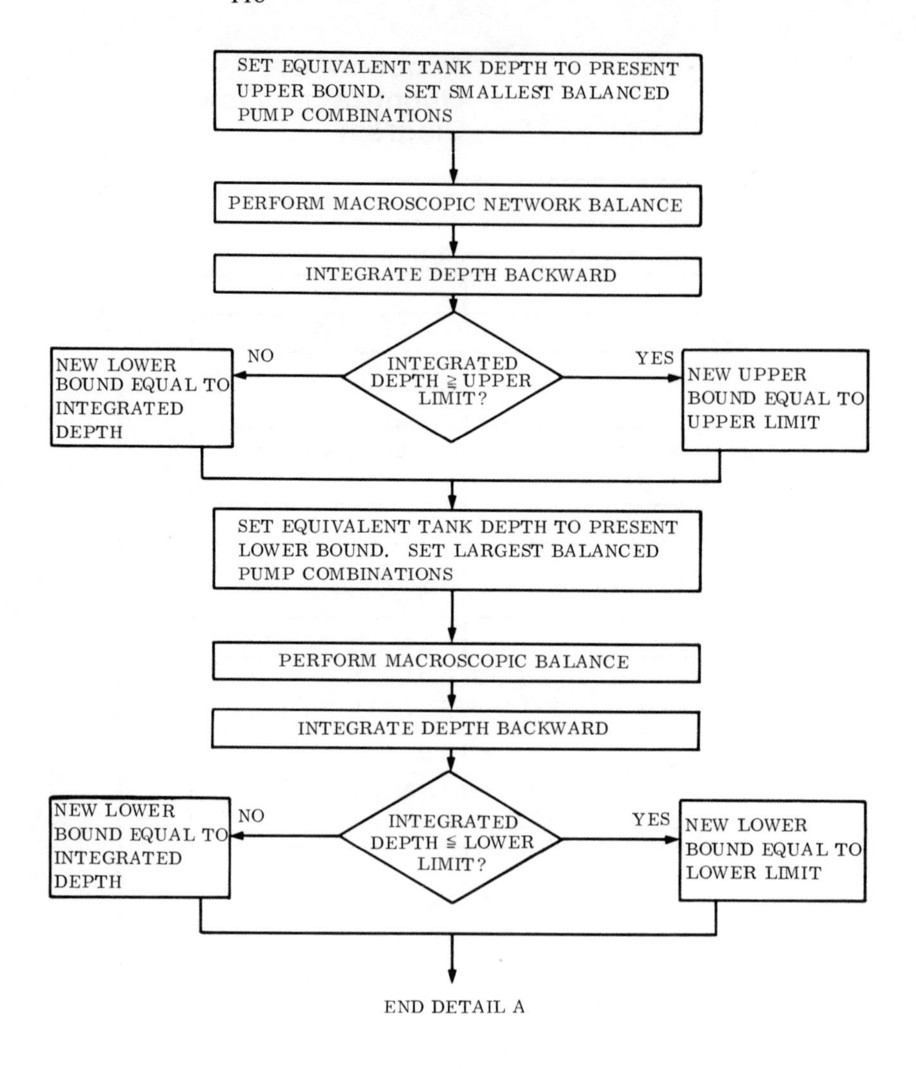

Figure B-2. Detail A: Calculate Equivalent Tank Depth Bounds

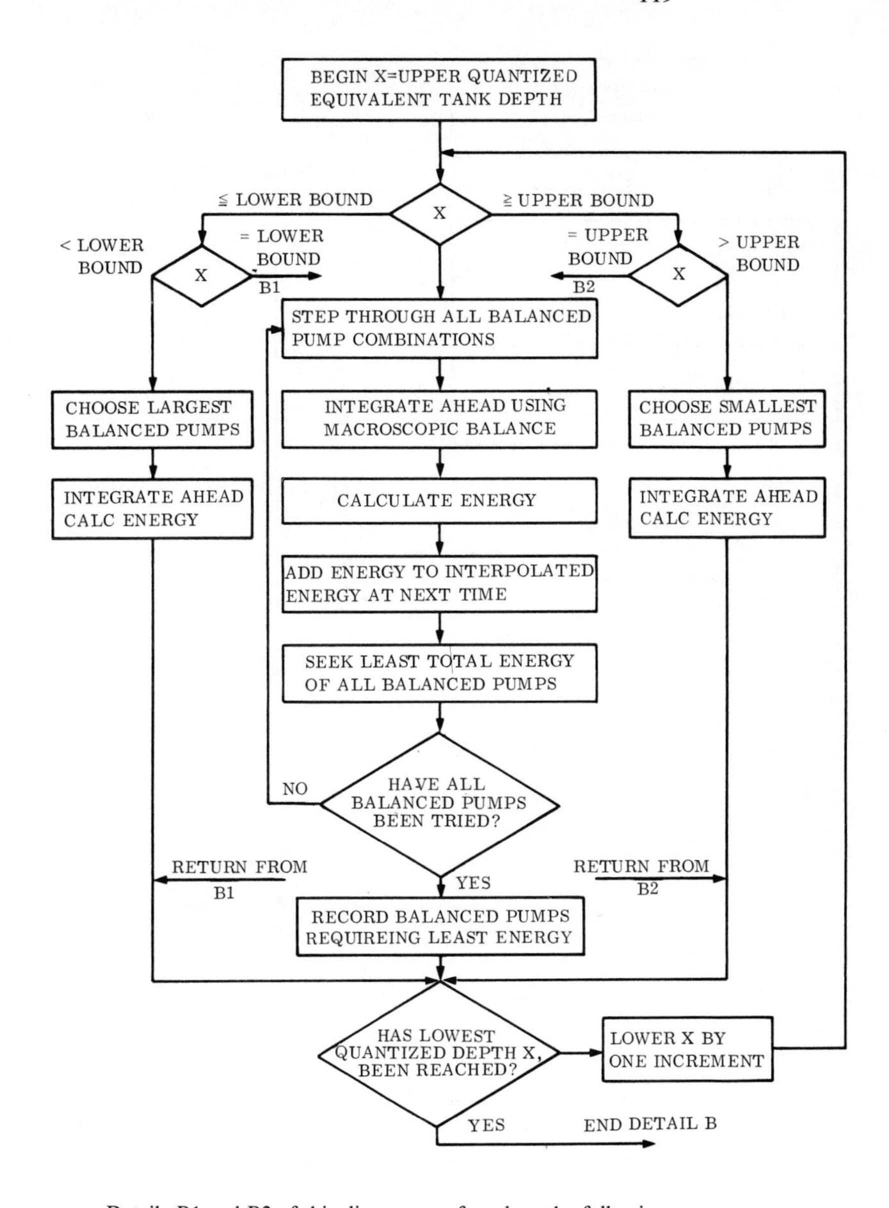

Details B1 and B2 of this diagram are found on the following pages.

Figure B-3. Detail B: DP Calculation at Each Stage in Time

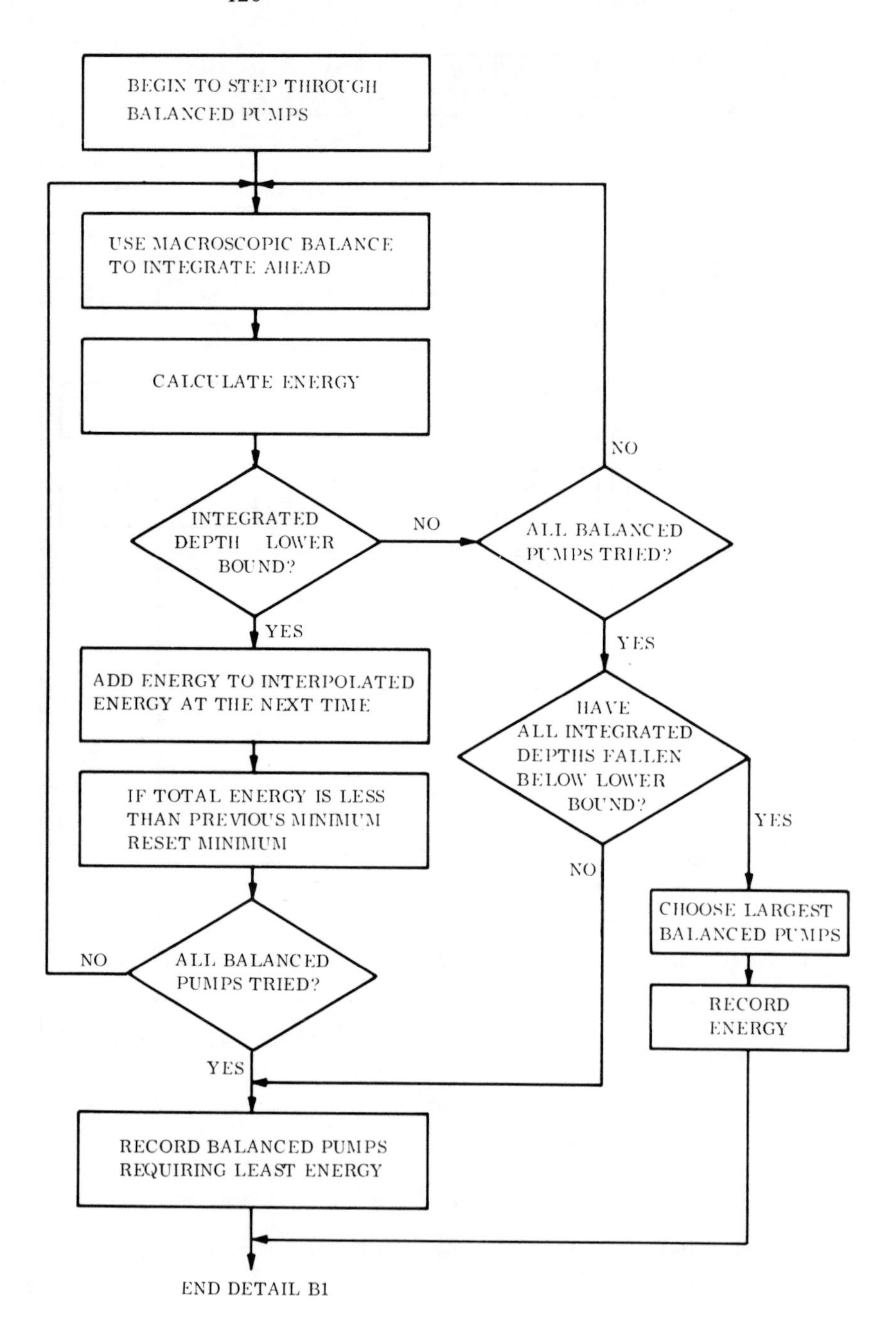

Figure B-4. Detail B1: Quantized State on Lower Bound

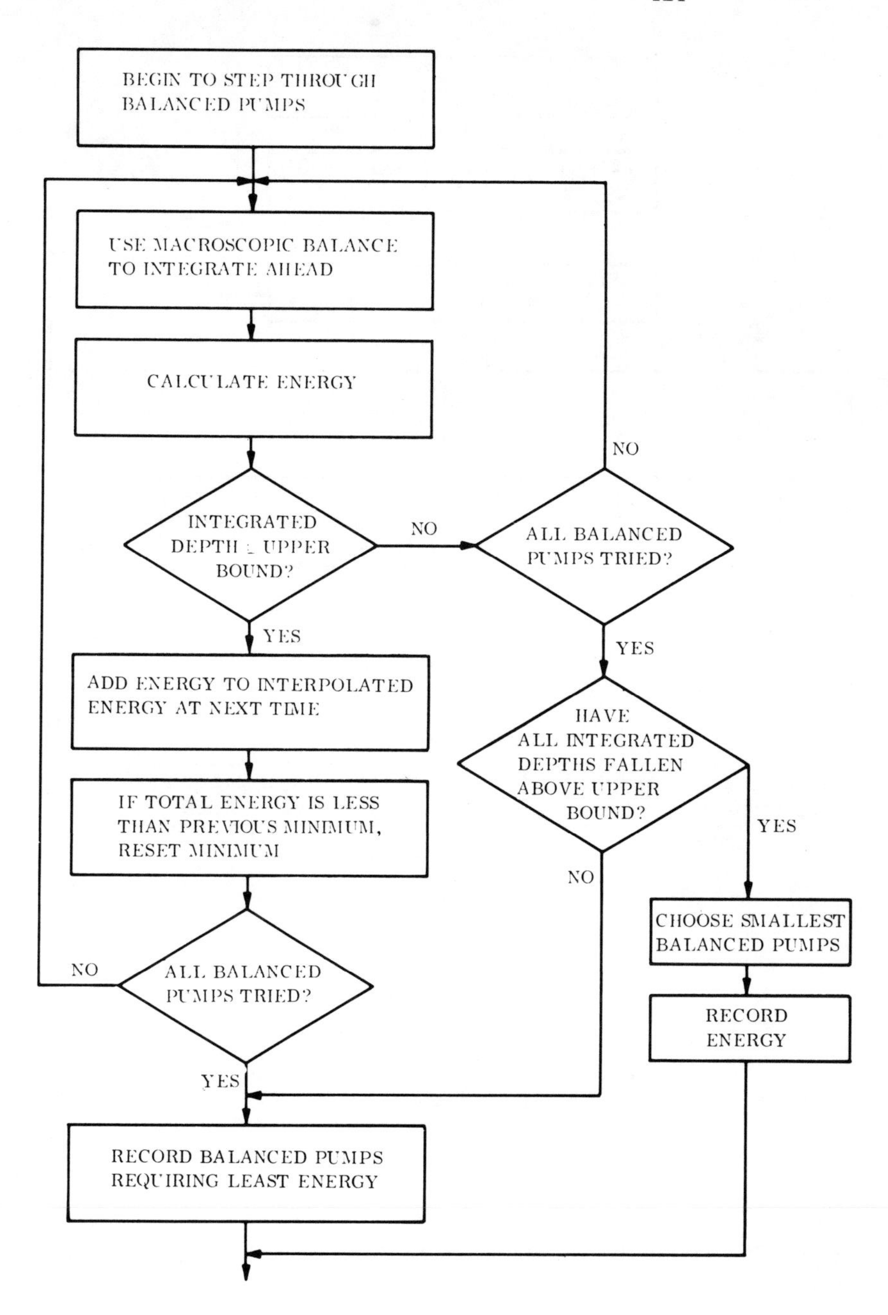

Figure B-5. Detail B2: Quantized State on Upper Bound

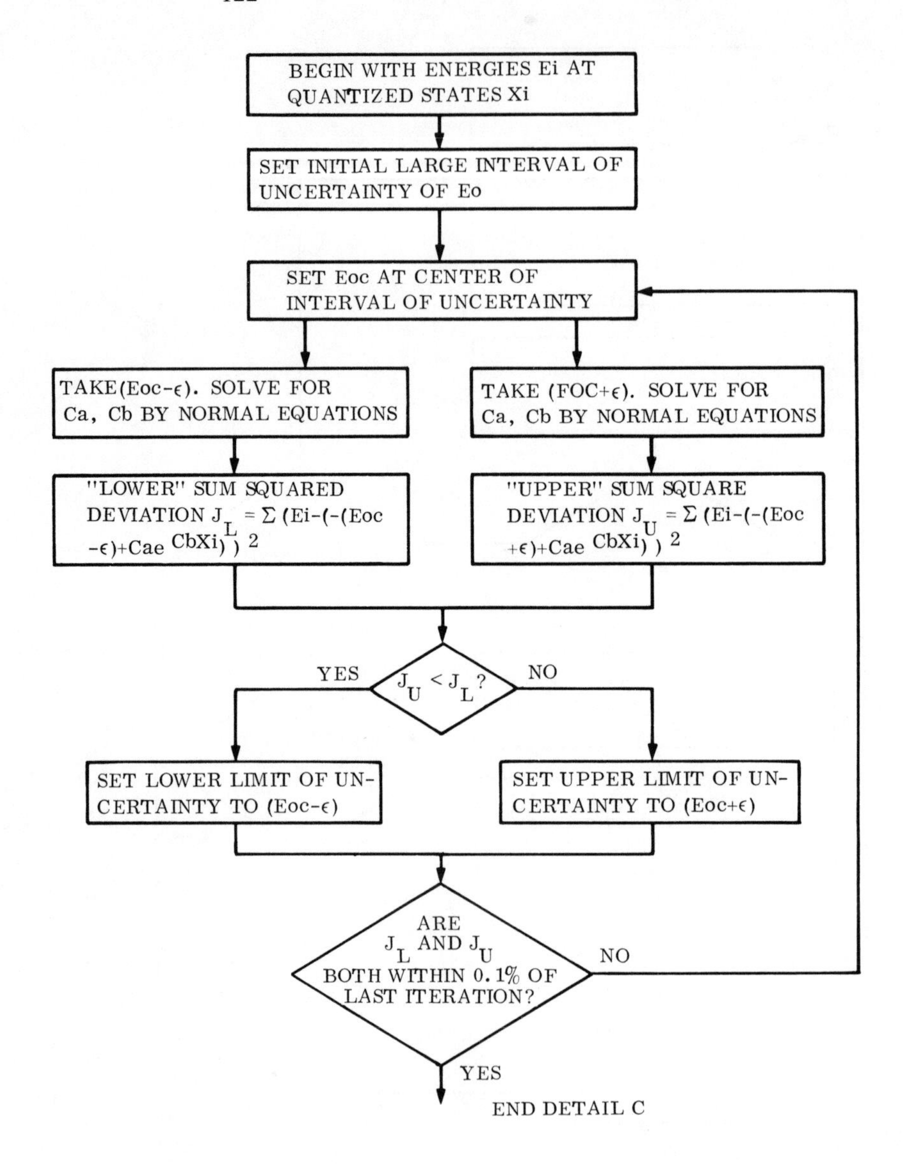

Figure B-6. Detail C: Evaluation of *Eo*, *Ca*, and *Cb* in Equation 5.19

Appendix C: Simulation Flow Diagram

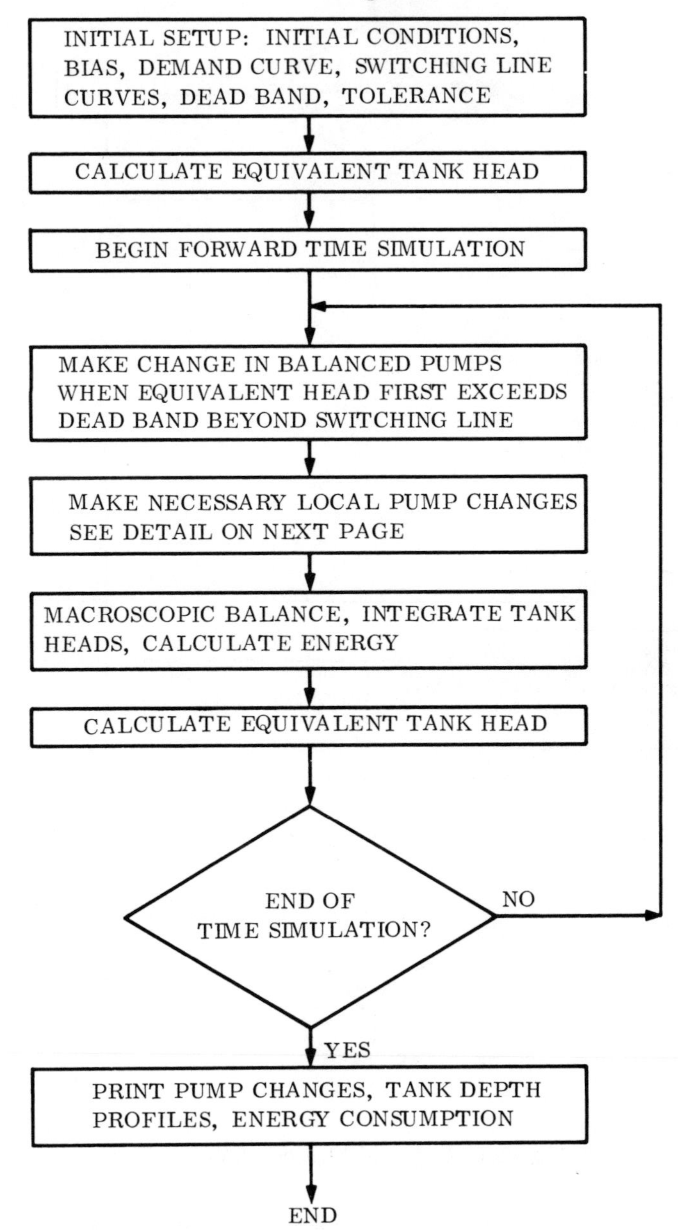

A detail of a local pump change is found on the following page.

Figure C-1. Flow Diagram of Simulated DP Control

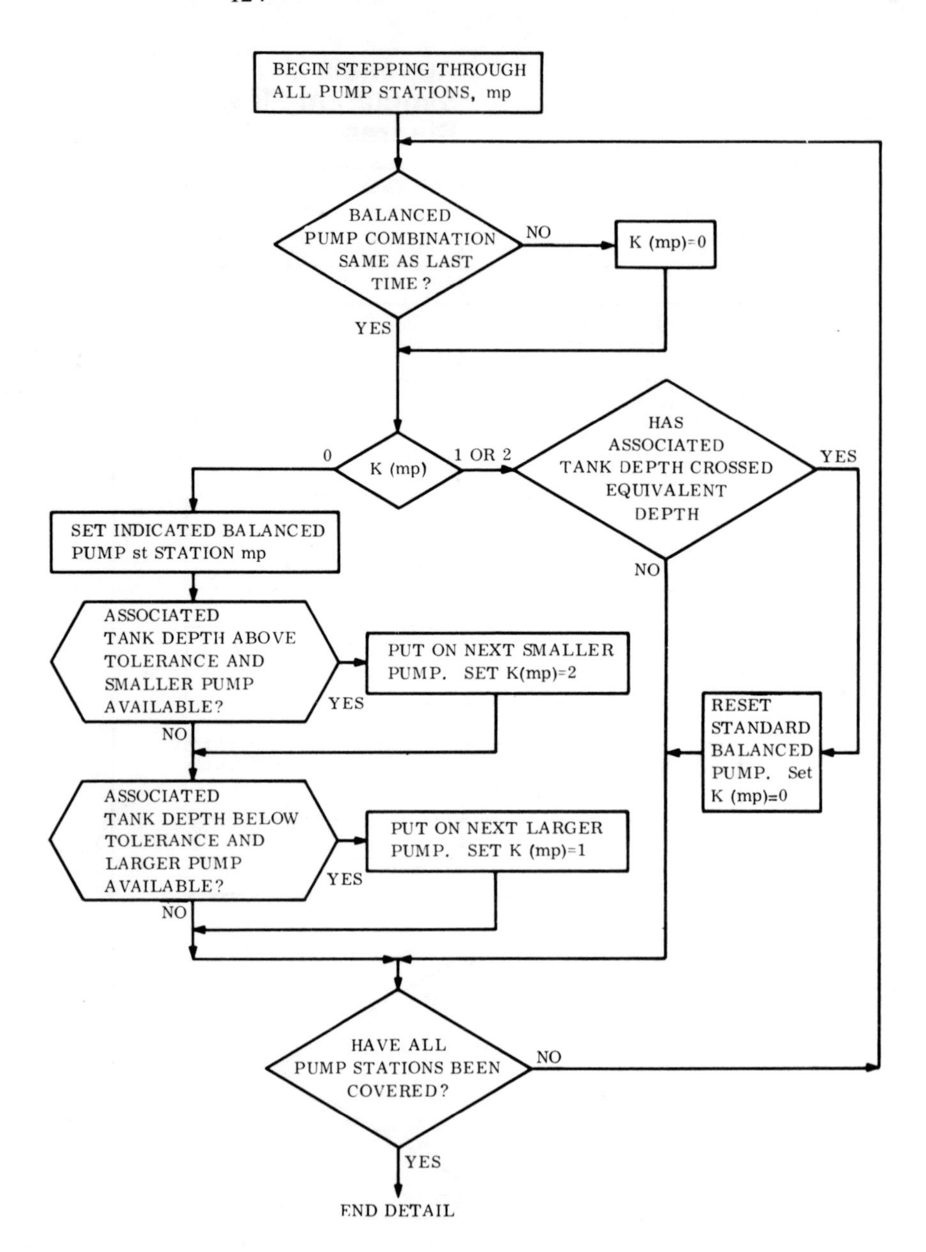

Figure C-2. Local Pump Change Detail

Appendix D: Week-Long Simulation

This appendix contains time plots of the two tank depths in the Torresdale-Fox Chase pressure district. Three curves are shown: "DP," which is the simulation of dynamic programming control; "FT," the simulation of floating tank control; and "As Run," the actual operating data (supervisory control). The simulations run from Monday, 10 January 1972, through Sunday, 16 January 1972 (see the section on "Dynamic Programming Control Characteristics" in Chapter 5 for a discussion of these plots).

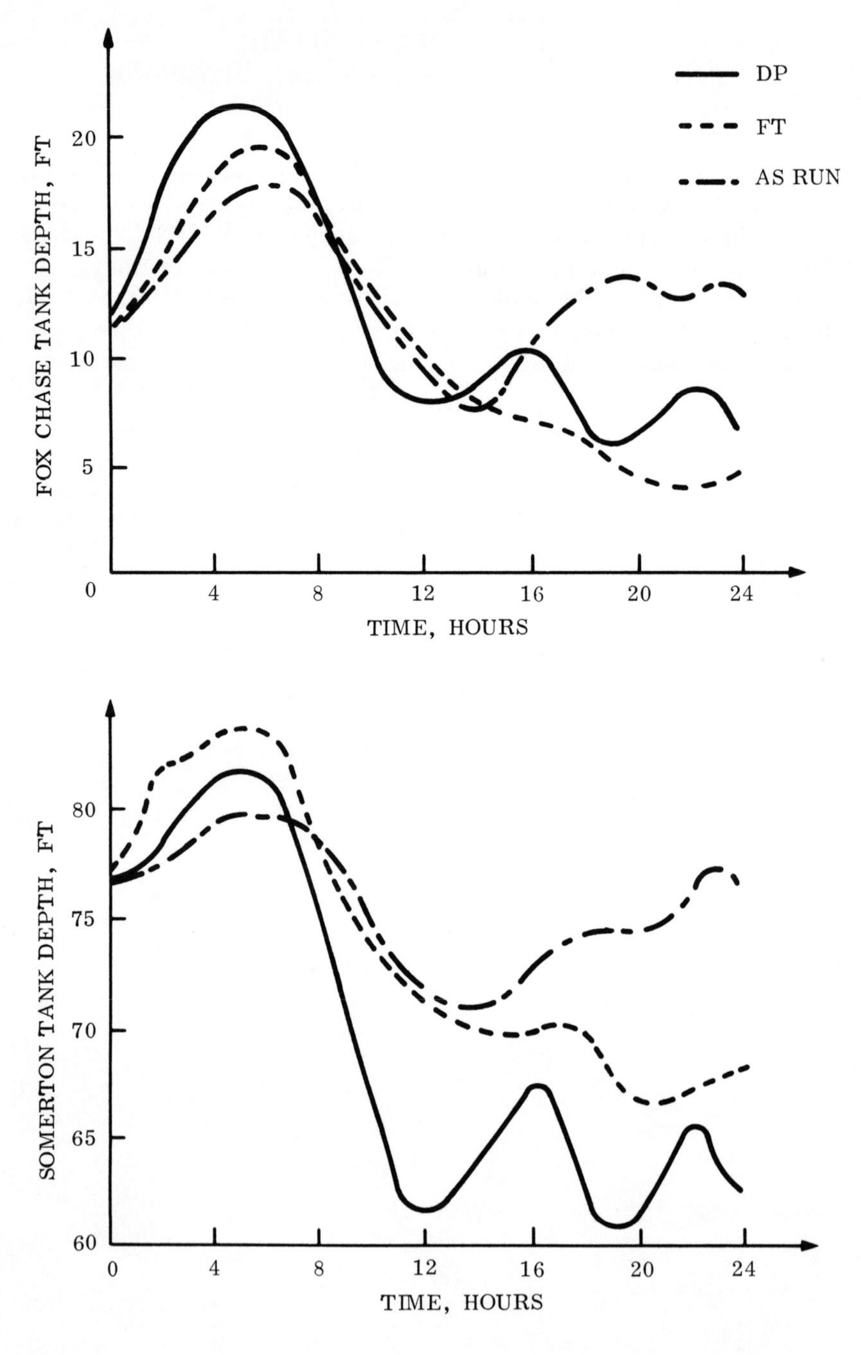

Figure D-1. Tank Depth Versus Time Plot for Monday (10 January 1972) Simulation

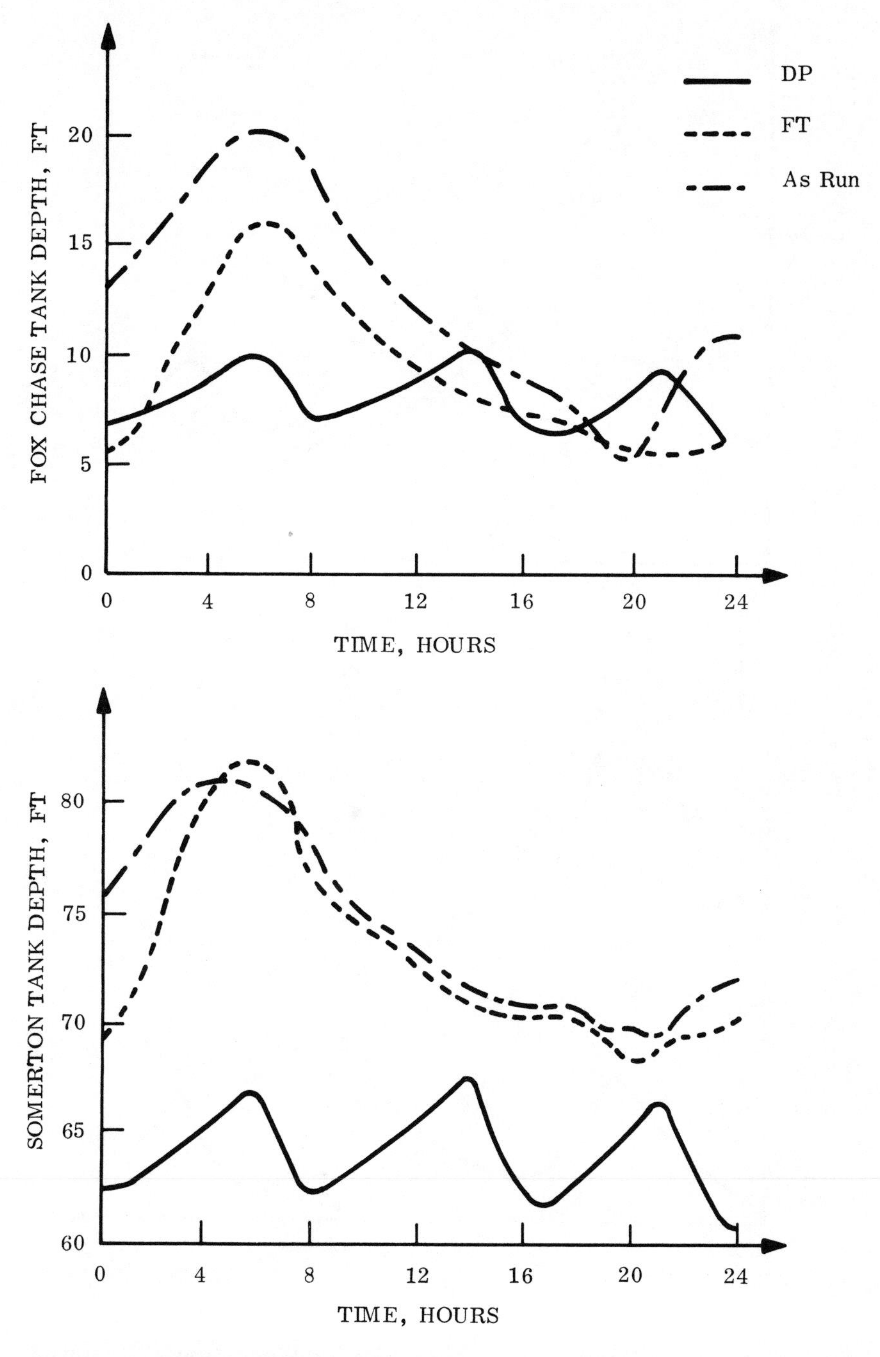

Figure D-2. Tank Depth Versus Time Plot for Tuesday (11 January 1972) Simulation

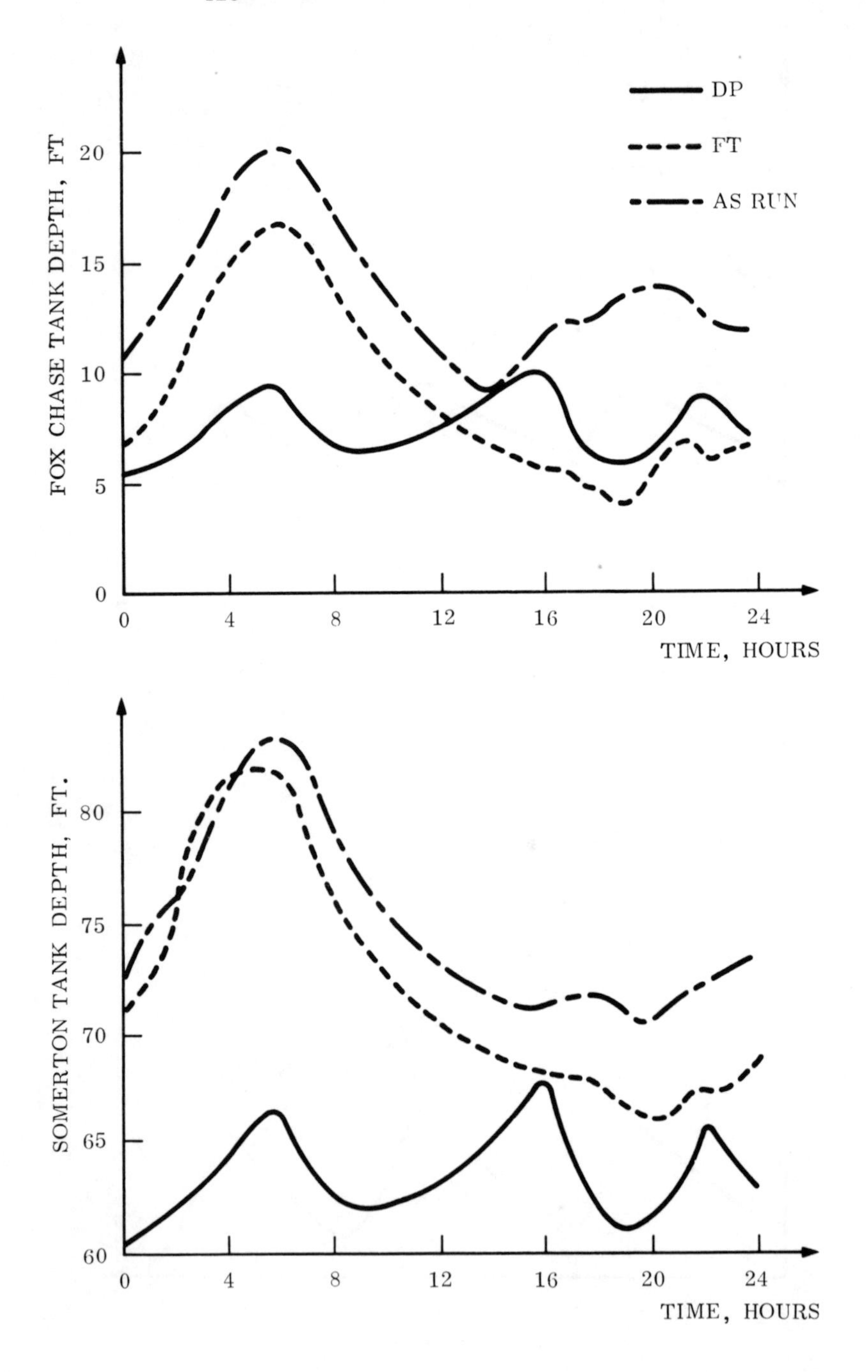

Figure D-3. Tank Depth Versus Time Plot for Wednesday (12 January 1972) Simulation

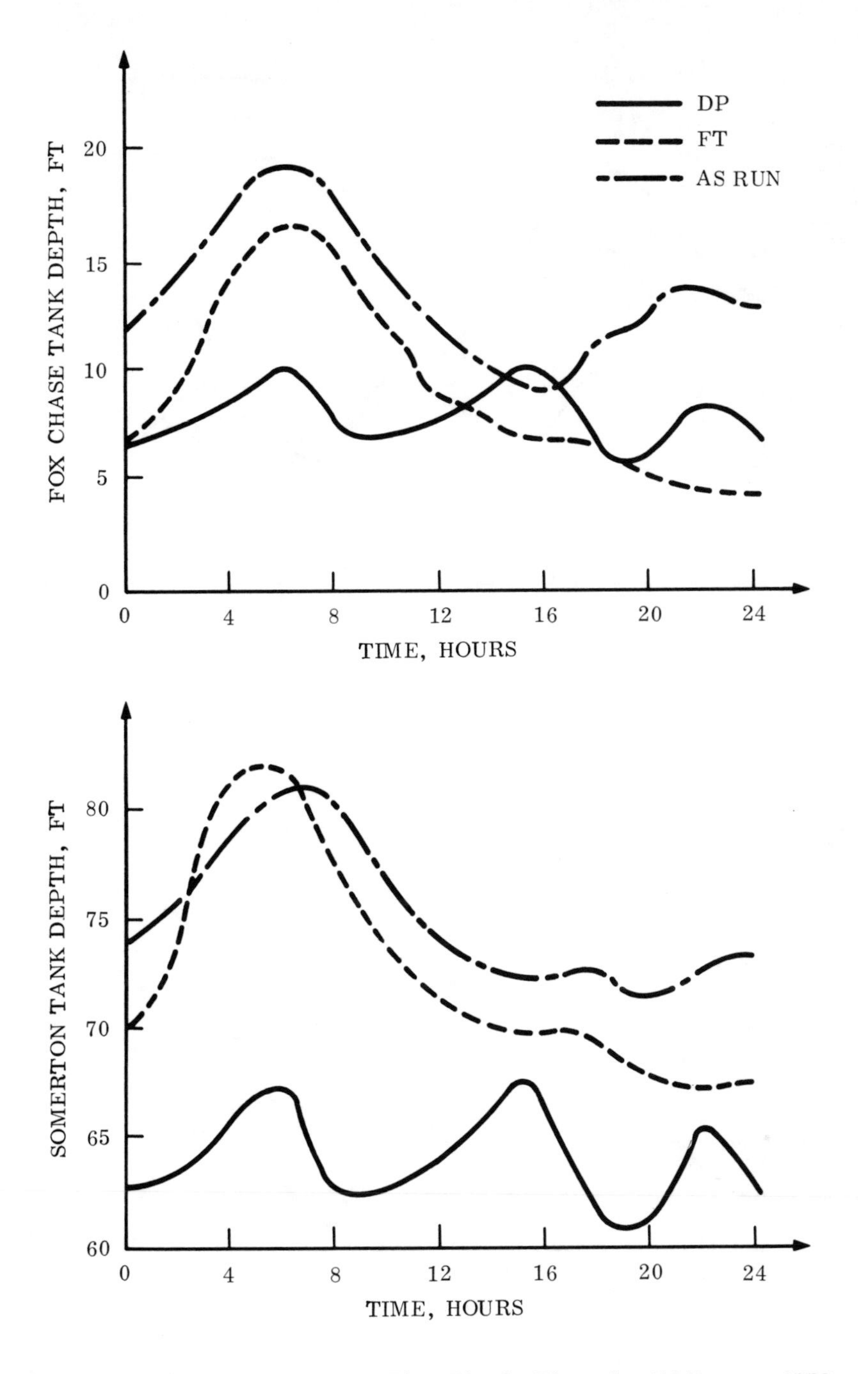

Figure D-4. Tank Depth Versus Time Plot for Thursday (13 January 1972) Simulation

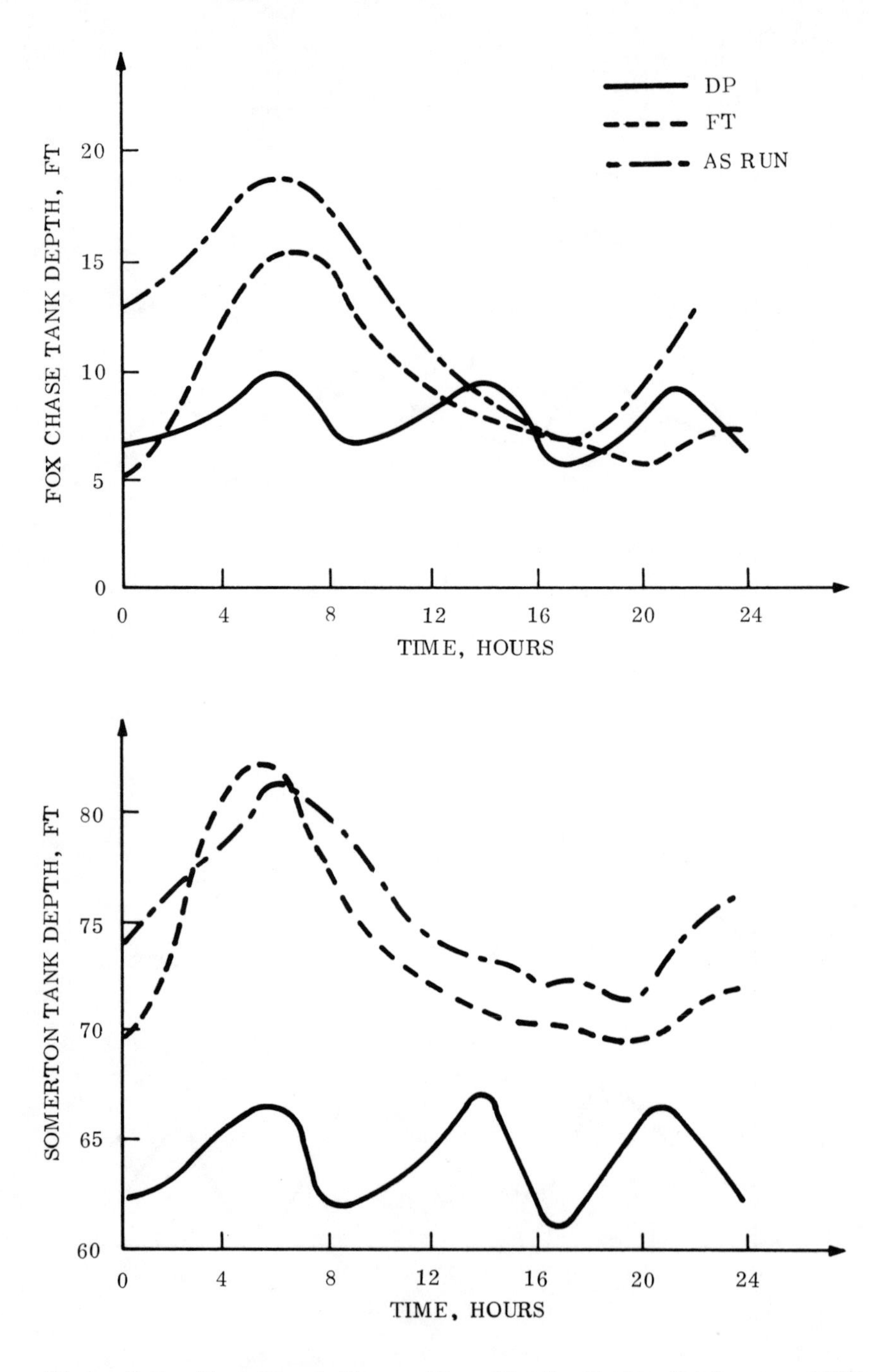

Figure D-5. Tank Depth Versus Time Plot for Friday (14 January 1972) Simulation

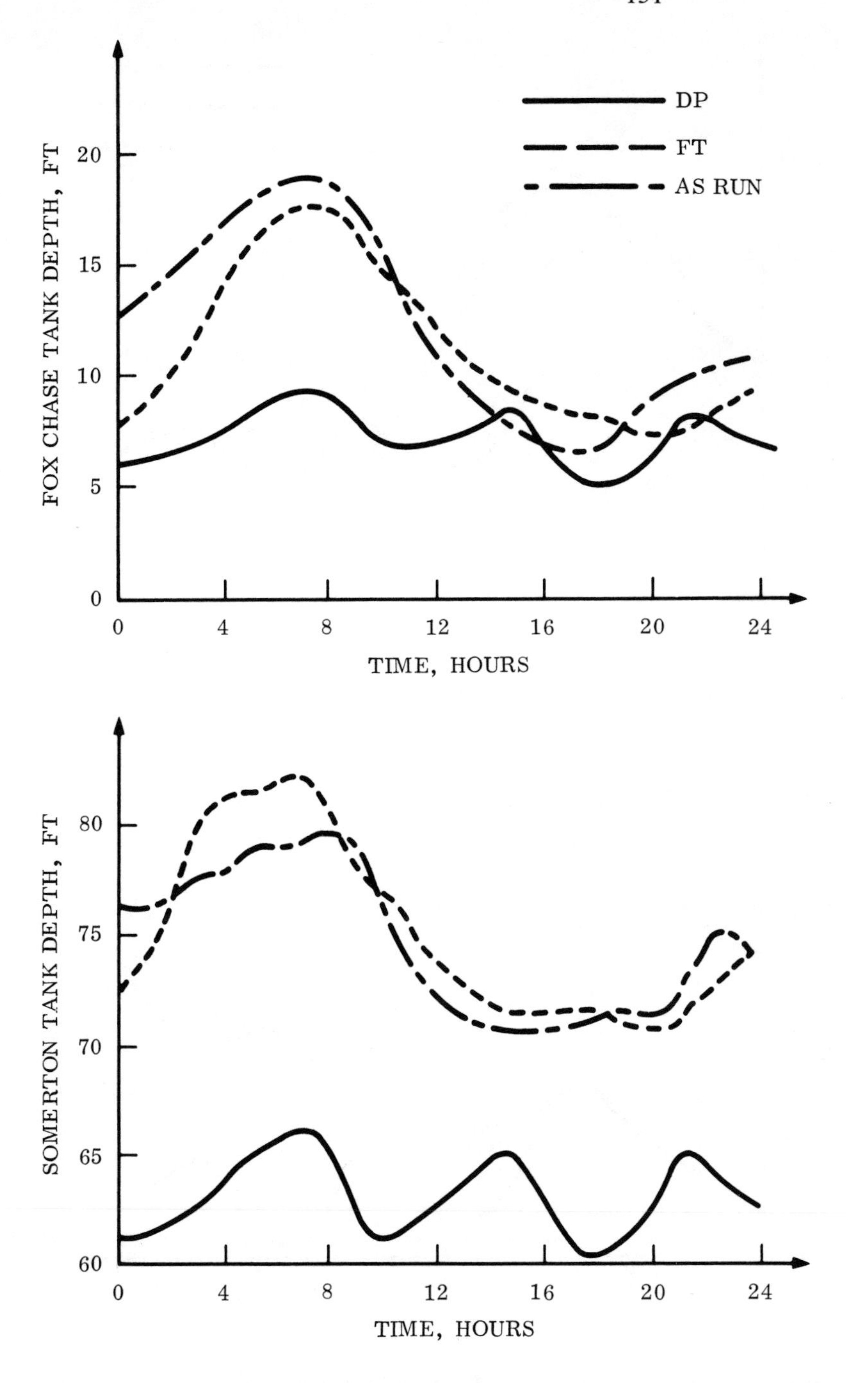

Figure D-6. Tank Depth Versus Time Plot for Saturday (15 January 1972) Simulation

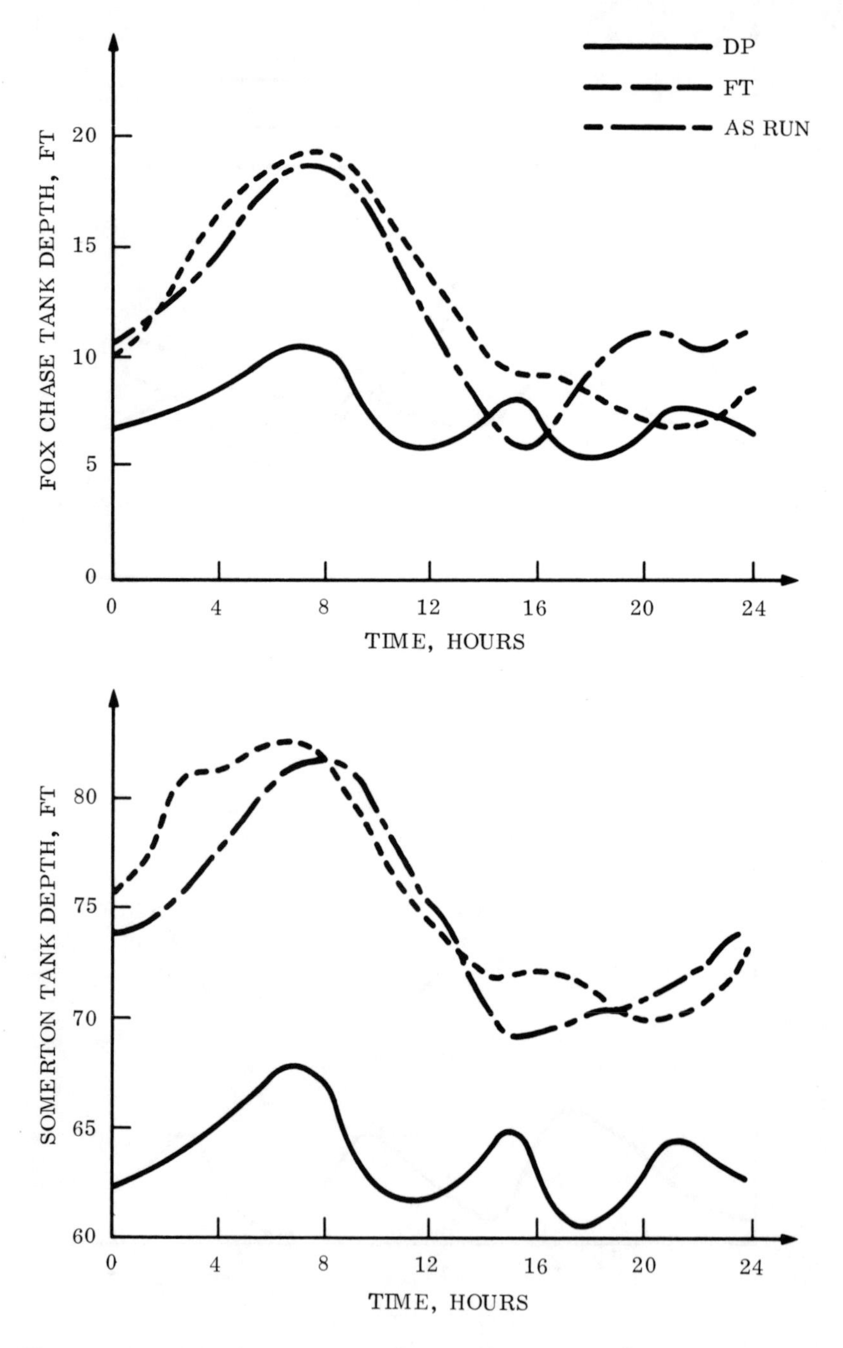

Figure D-7. Tank Depth Versus Time Plot for Sunday (16 January 1972) Simulation

Bibliography

[1] Baxter, S.S. "Use of Computers in Water Utility Work At Philadelphia," *Journal of The American Water Works Association 58, (J. AWWA)* No. 8 (August 1966), pp. 1011-1020.

[2] Beck, R.W., and Associates. "Computer System for Automatic Control and Commerical and Engineering Data Processing," City and County of Honolulu, Hawaii, (February 1972).

[3] Bellman, R.E., and S.E. Dreyfus. *Applied Dynamic Programming*. Princeton, N.J.: Princeton University Press, 1972.

[4] Box, G.E., and G.M. Jenkins, *Time Series Analysis Forecasting and Control*. San Francisco: Holden-Day, 1970.

[5] Brock, D.A. "Closed-Loop Automatic Control of Water System Operations," *J. AWWA, 55,* No. 4 (April 1963), pp. 467-480.

[6] Brock, D.A. "Determination of Optimum Storage in Distribution System Design," *J. AWWA, 55,* No. 8 (August 1963), pp. 1027-1036.

[7] Brock, D.A. Supervisor, Operations Research and Analysis, City of Dallas Deartment of Water Works, Dallas, Tex. (personal communication).

[8] Bryson, A.E., and Y. Ho. *Applied Optional Control*. Waltham, Mass.: Ginn, 1969.

[9] Buras, N. "Conjunctive Operation of Dams and Aquifers," *Journal of the Hydraulics Division, Proceedings of the American Society of Civil Engineers (Proc. ASCE) 89,* No. NY6 (November 1963), pp. 111-131.

[10] Buras, N., and Z. Schweig. "Aqueduct Route Optimization by Dynamic Programming," *Journal of the Hydraulics Division, Proc. ASCE, 95,* No. HY5 (September 1969), pp. 1615-1631.

[11] Butcher, W.S., Y.Y. Haimes, and W.A. Hall. "Dynamic Programming for the Optimal Sequencing of Water Supply Projects," *Water Resources Research, 5,* No. 6 (December 1969), pp. 1196-1204.

[12] Camp, T.R., and J.C. Lawler. "Water Distribution," *Handbook of Applied Hydraulics*. New York: McGraw-Hill, 1969, pp. 37.1-37.21.

[13] Carlson, C.E. "The Denver System of Water Works Controls," *J. AWWA, 63,* No. 8, (August 1971), pp. 513-516.

[14] Cross, H. "Analysis of Flow in Networks of Conduits or Conductors," University of Illinois Engineering Experimental Station, Urbana, Ill., *Bulletin No. 286, (1936)*.

[15] Davis, M.L., and M.H. Diskin. "A Study of Alternative Expressions Describing Generalized Distribution Network Head Loss Charac-

teristics,'' Appendix II of ''A Study of the Applicability of Generalized Distribution Network Head Loss Characterisitcis,'' Department of Civil Engineering, University of Illinois, Urbana, Ill., (June 1965).

[16] Deb, A.K., and A.K. Sarkar. ''Optimization in Design of Hydraulic Network,'' *Journal of the Sanitary Engineering Division, Proc. ASCE, 97,* No. SA2 (April 1971), pp. 141-159.

[17] Debs, A.S., and R.E. Larson. ''A Dynamic Estimator for Tracking the State of a Power System,'' *IEEE Transactions on Power Apparatus and Systems, Pas-89,* No. 7 (September/October 1970), pp. 1670-1682.

[18] DeMoyer, R., Jr., H.D. Gilman, and M.Y. Goodman. ''Dynamic Computer Simulation and Control Methods for Water Distribution Systems,'' General Electric Document No. 73SD205, prepared for U.S. Department of Interior Office of Water Resources Research, Project C-3166, Philadelphia, Pa. (1973).

[19] de Nuefville, R., J. Schaake, Jr., and J.H. Stafford. ''Systems Analysis of Water Distribution Networks,'' *Journal of the Sanitary Engineering Division, Proc. ASCE, 97,* No. SA6 (December 1971), pp. 825-842.

[20] Epp, R., and A.G. Fowler. ''Efficient Code for Steady-State Flows in Networks,'' *Journal of the Hydraulics Division, Proc. ASCE, 96,* No. HY1 (January 1970), pp. 43-56.

[21] Fair, G.M., and Geyer, J.C. ''Distribution of Water,'' *Water Supply and Waste-Water Disposal.* New York: Wiley, Chap. 13, pp. 336-368.

[22] Frenz, C.M. ''Automated System Control,'' *J. AWWA, 63,* No. 8, (August 1971), pp. 508-512.

[23] Gibson, J.E., *Nonlinear Automatic Control.* New York: MacGraw-Hill, 1963.

[24] Gilman, H.D., M.Y. Goodman, and R.V. Metkowski. ''Mathematical Modeling of Water Distribution Systems,'' General Electric Document No. 71SD241, Philadelphia, Pa.

[25] Gilman, H.D., M.Y. Goodman, R. DeMoyer, Jr., and R.V. Radziul. ''Closed Loop Control is Coming,'' *Water and Wastes Engineering,* Vol. 10, No. 6, (June 1973), pp. 42-44.

[26] Gilman, H.D., M.Y. Goodman, and R. DeMoyer, Jr. ''Replication Modeling for Water-Distribution Control,'' *J. AWWA,* Vol. 65, No. 4, (April 1973), pp. 255-260.

[27] Gracie, C. ''Analysis of Distribution Demand Variations,'' *J. AWWA, 58,* No. 1 (January 1966), pp. 51-66.

[28] Graeser, H.J. "Use of Computers in Water Utility Work at Dallas," *J. AWWA, 58,* No. 8 (August 1966), pp. 1023-1037.

[29] Hall, W.A., W.S. Butcher, and A. Esogbue. "Optimization of the Operation of a Multiple-Purpose Reservoir by Dynamic Programming," *Water Resources Research, 4,* No. 3 (June 1968), pp. 471-477.

[30] Hanke, S.H. "Demand for Water under Dynamic Conditions," *Water Resources Research, 6,* No. 5 (October 1970), pp. 1253-1261.

[31] Harleman, D.R.F. "Flow Resistance in Pipes and Fittings," *Water and Sewage Works* (November 1950), pp. 465-471.

[32] Himmelblau, D.M. *Process Analysis by Statistical Methods.* New York: Wiley, 1970.

[33] Hudson, W.D. President, Pitometer Associates, Chicago, Ill. (personal communication).

[34] Jacoby, S. L. S. "Design of Optimal Hydraulic Networks," *Journal of the Hydraulics Division, Proc. ASCE, 94,* No. HY3, (May 1968), pp. 641-661.

[35] Jenkins, G.M., and D.G. Watts. *Spectral Analysis and its Applications.* San Francisco: Holden-Day, 1969.

[36] Kally, E. "Pipeline Planning by Dynamic Computer Programming," *J. AWWA, 61,* No. 3 (March 1969), pp. 114-118.

[37] Kesavan, H.K., and M. Chandrashekar. "Graph-Theoretic Models for Pipe Network Analysis," *Journal of the Hydraulics Division, Proc. ASCE, 98,* No. HY2 (February 1972), pp. 345-364.

[38] Kuranz, J.H. "Automatic of Water Utilities: Wauskesba, Wisconsin," *J. AWWA, 59,* No. 1 (January 1967), pp. 93-99.

[39] Lai, F., and J.C. Schaake, Jr. "A Model for Capacity Expansion Planning of Water Distribution Networks," Water Research Center University of Massachusetts, Amherst, Report No. 131, (October 1970).

[40] Lam, C.F., and M.L. Wolla. "Computer Analysis of Water Distribution Systems: Part I—Formulation of Equations," *Journal of the Hydraulics Division, Proc. ASCE, 98,* No. HY2 (February 1972), pp. 335-343.

[41] Lam, C.F., and M.L. Wolla "Computer Analysis of Water Distribution Systems: Part II—Numerical Solution," *Journal of the Hydraulics Division, Proc. ASCE, 98,* No. HY3 (March 1972), pp. 447-460.

[42] Larson, R.E. "A Survey of Dynamic Programming Computational Precedures," *IEEE Transactions on Automatic Control* (December 1967), pp. 767-774.

[43] Larson, R.E., W.F. Tinney, and J. Peschon. "State Estimation in Power Systems: Part I—Theory and Feasibility," *IEEE Transactions On Power Apparatus and Systems, Pas-89,* No. 3 (March 1970), pp. 345-352.

[44] Linaweaver, F.P., Jr., J.C. Geyer, and J.B. Wolff, "Summary Report on the Residential Water Use Research Project," *J. AWWA, 59,* No. 3 (March 1967), pp. 267-282.

[45] Machis, A. "Computers in Distribution Analysis," *J. AWWA, 60,* No. 6 (June 1968), pp. 634-639.

[46] Martin, D.W., and G. Peters. "The Application of Newton's Method to Network Analysis by Digital Computer," *Journal of Institute of Water Engineers, 17* (1963), pp. 115-129.

[47] Matucha, J. "Don't Reject Dynamic Programming for Complex Systems," *Computer Decisions* (April 1972), pp. 18-23.

[48] McCausland, I. *Introduction to Optimal Control.* New York: Wiley, 1969.

[49] McCormick, M., and C.J. Bellamy. "A Computer Program for the Analysis of Networks of Pipes and Pumps," *Journal of the Institution of Engineers (Australia), 40* (March 1968), pp. 51-58.

[50] McIlroy, M.S. "Direct Reading Electric Analyzer for Pipeline Networks," *J. AWWA, 42* (April 1950), p. 347.

[51] McPherson, M.B. "Generalized Distribution Network Head-Loss Characteristics," *Transactions ASCE, 126* (1961), pp. 1190-1234.

[52] ——. "Ground Storage Booster Pumping Hydraulics," *J. AWWA, 58,* No. 3 (March 1966), pp. 311-325.

[53] ——. Chairman, AWWA Research Committee on Distribution Systems, Marblehead, Mass. (personal communication).

[54] ——. "A Study of the Applicability of Generalized Distribution Network Head Loss Characteristics," Department of Civil Engineering, University of Illinois, Urbana, Ill., (June 1965).

[55] ——. "Water Distribution Research and Applied Development Needs," prepared for review by AWWA Research Committee on Distribution Systems, AWWA Annual Conference, Chicago, (June 5 1972), pp. 1-15.

[56] McPherson, M.B., and M. Heidari. "Power Consumption with Elevated Storage Compared to Direct and Booster Pumping," *J. AWWA, 58,* No. 12 (December 1966), pp. 1585-1594.

[57] McPherson, M.B., and R. Prasad. "Distribution System Equalizing Storage Hydraulics," *Journal of the Hydraulics Division, Proc. ASCE, 92,* No. HY6 (November 1966), pp. 151-177.

[58] McPherson, M.B., and R. Prasad. "Power Consumption for Equalizing-Storage Operating Options," *J. AWWA, 58,* No. 1 (Janaury 1966), pp. 67-90.

[59] McPherson, M.B., and G. Wood. "Operating Options for Pumped Equalizing Storage," *J. AWWA, 57,* No. 7 (July 1965), pp. 869-884.

[60] Mehra, R.K. "On the Identification of Variances and Adaptive Kalman Filtering," *IEEE Transactions on Automatic Control, AC-15,* No. 2 (April 1970), pp. 175-184.

[61] Meier, W.L., Jr., and C.S. Beightler. "An Optimization Method for Branching Multistage Water Resource Systems," *Water Resources Research, 3,* No. 3 (1967), pp. 645-652.

[62] Meyer, J.M., Jr., and G.F. Mangan. "System Pinpoints Urban Water Needs," *Environmental Science & Technology, 3,* No. 10 (October 1969), pp. 904-911.

[63] Michel, H.L., and J.P. Wolfner. "Know What's Happening in Your Water System," *The American City* (June 1972).

[64] Neel, R.C. "Computer Applications in Distribution," *J. AWWA, 63,* No. 8 (August 1971), pp. 485-489.

[65] Nemhauser, G.L. *Introduction to Dynamic Programming.* New York: Wiley, 1966.

[66] City of Philadelphia, *Water Department Biennial Report, 1968-1969.* pp. 5-44.

[67] Poertner, H.G. "Existing Automation, Control and Intelligence Systems for Metropolitan Water Facilities," Department of Civil Engineering, Colorado State, University, Fort Collins, Colo., Metropolitan Water Intelligence Systems Project, Technical Report No. 1.

[68] Radziul, J.V. "Automation and Instrumentation Committee's Purpose," *J. AWWA, 63,* No. 8 (August 1971), pp. 467-469.

[69] Radziul, J.V. Chief, Research and Development, Philadelphia Water Department, Philadelphia, Pa. (personal communication).

[70] Reh, C.W. "Hydraulics of Water Distribution Systems," *Proceedings, Fourth Sanitary Engineering Conference: Water Distribution Systems,* Urbana, Ill. (February 1962), pp. 6-22.

[71] Reynolds, R.R., and W.R. Madsen. "Automation in California's State Water Project," *Journal of the Pipeline Division, Proc. ASCE, 93,* No. PL3 (November 1967), pp. 15-23.

[72] Salas-LaCruz, J.D., and V. Yevjevich. "Stochastic Structure of Water Use Time Series," Hydrology Papers, Colorado State University, Fort Collins, Colo. No. 52 (June 1972).

[73] Shamir, U., and C.D.D. Howard. "Water Distribution Systems

Analysis," *Journal of the Hydraulics Division, Proc. ASCE, 94,* No. HY1 (January 1968), pp. 219-234.

[74] Shirley, W., and J.J. Bailey. "Use of Digital Computers in Distribution System Analyses," *J. AWWA, 58,* No. 12 (December 1966), pp. 1575-1584.

[75] Smith, G.H. "Central Surveillance and Control of the City of Houston Water System," presented at the *AWWA* Conference, Chicago (June 1972).

[76] Sorenson, H.W. "Kalman Filtering Techniques," *Advances in Control Systems,* ed. C.T. Leondes. New York: Academic Press, 1966, Vol. 3 pp. 219-292.

[77] Sorenson, H.W. "Least-Squares Estimation: from Gauss to Kalman," *IEEE Spectrum* (July 1970), pp. 63-68.

[78] Stephenson, D. "Operations Research Techniques for Planning and Operation of Pipeline Systems," presented at The Pipeline Engineering Convention, (1970).

[79] Sturman, G.M. "Systems Analysis for Urban Water Supply and Distribution," *Journal of Environmental Systems, 1,* No. 1 (March 1971), pp. 67-76.

[80] Sturman, G.M., P.H. Gilbert, and J.P. Wolfner, "Systems Engineering for Urban Utilities," *J. AWWA, 63,* No. 9 (September 1971), pp. 565-570.

[81] Tolle, W.A. Chief, Commercial Division, Board of Water Commissioners, Denver, Colo. (personal communication).

[82] Van Dyke, R.P. "Computerized Operation of Distribution Systems," *J. AWWA, 62,* No. 4 (April 1970), pp. 234-240.

[83] Voyles, C.F., and H.R. Wilke. "Selection of Circuit Arrangements for Distribution Network Analysis by the Hardy Cross Method," *J. AWWA, 54,* No. 3 (March 1962), pp. 285-290.

[84] Wilde, D.J. *Optimum Seeking Methods.* Englewood Cliffs, N.J.: Prentice-Hall, 1964.

[85] Wiseman, R.A. "A Demonstration of Solution Uniqueness in Network Analysis," Appendix I of "A Study of the Applicability of Generalized Distribution Network Head Loss Characteristics," Department of Civil Engineering, University of Illinois, Urbana, Ill. (June 1965).

[86] Wislicenus, G.F. "Centrifugal Pumps," *Marks Mechanical Engineering Handbook.* New York: McGraw-Hill, 1951, pp. 1848-1865.

[87] Wolff, J.B. "Peak Demand in Residential Areas," *J. AWWA, 53,* No. 10 (October 1961), pp. 1251-1260.

[88] Wong, P.J., and R.E. Larson. ''Optimization of Natural-Gas Pipeline Systems Via Dynamic Programming,'' *IEEE Transactions on Automatic Control, AC-13,* No. 5 (October 1968), pp. 475-481.

[89] Wood, D.J. ''Analog Analysis of Water Distribution Networks,'' *Transportation Engineering Journal, Proc. ASCE, 97,* No. TE2 (May 1971), pp. 281-290.

[90] Wood, D.J., and C.O.A. Charles. ''Hydraulic Network Analysis Using Linear Theory,'' *Journal of the Hydraulics Division, Proc. ASCE, 98,* No. HY7 (July 1972), pp. 9031-1170.

[91] Young, G.K., Jr. ''Finding Reservoir Operating Rules,'' *Journal of the Hydraulics Division, Proc. ASCE, 93,* No. HY6 (November 1967), pp. 297-319.

[92] Zarghamee, M.S., ''Mathematical Model for Water Distribution Systems,'' *Journal of the Hydraulics Division, Proc. ASCE, 97,* No. HY1 (January 1971), pp. 1-14.

Index

Index

Index

About the Authors

Robert DeMoyer, Jr. joined the General Electric Company in 1966, and has held positions in several departments since that time. He is with the Re-entry and Environmental Systems Department which, in an effort to find new applications of aerospace technology, has conducted several studies and experiments in water distribution. Dr. DeMoyer, who received the B.S. degree from Lehigh University, and the M.S. and Ph.D. degrees from the Polytechnic Institute of New York is a professional engineer in the state of Pennsylvania, and has coauthored several journal articles in the area of water distribution modeling and control.

Lawrence B. Horwitz is an associate professor of electrical and system engineering at the Polytechnic Institute of New York, and is a registered professional engineer in the state of New York. Prior to this appointment he was a member of the technical staff at the Bell Telephone Laboratories. He received the B.E. degree from the City College of New York, and the M.S. and Ph.D. degrees from New York University. The author of several technical papers and reports, Dr. Horwitz specializes in the methodologies and applications of System Science and Engineering and has served as an industrial consultant in these areas.